Stone, Log and Earth Houses

Stone, Log and Earth Houses

Building with elemental materials

Magnus Berglund

The Taunton Press

**Cover and text photos by Magnus Berglund,
except where noted.
Right-hand cover photo by Lyle Hymer-Thompson.**

Portions of the material in Chapter 2 were originally
published in *Fine Homebuilding* magazine.

First printing: November 1986
International Standard Book Number: 0-918804-61-2
Library of Congress Catalog Card Number: 86-50403
Printed in the United States of America

A FINE HOMEBUILDING Book

FINE HOMEBUILDING® is a trademark of The Taunton Press, Inc.,
registered in the U.S. Patent and Trademark Office.

The Taunton Press, Inc.
63 South Main Street
Newtown, Connecticut 06470

Acknowledgments

Building a life is like building a house, it's no fun to do it by yourself. This book is dedicated to the helpers in my life—Cathleen, Carma and Eben.

I would like to thank the following people who contributed both directly and indirectly to the creation of this book: my editor, Laura Cehanowicz Tringali, who believed in this project and helped to define it; the late Ken Kern, the guru of owner-builders and treasured friend; Fred Vinson, who guided me to the opportunities in remodeling; David Easton, who caught my imagination with his rammed-earth missionary zeal; and all those whose houses are pictured in this book. Most of them spent hours answering my questions, cleaning up for photographs, digging out negatives and drawing plans. I value their support and their friendship. These builders and owners embody a resourcefulness and generosity of spirit that is apparent in their homes and lives, and they deserve the real credit for whatever success this book might attain.

Contents

Prologue

This is a dream book. On the surface it is a book about building with three types of natural, often neglected, materials, but underneath I have threaded a subtle undercurrent that can catch you unaware. There are dreams here that are part of the fabric of our culture, the collective unconscious of shelter. There are wispy glimpses of stone castles on the cliffs above the sea. There are hints of log houses under postcard skies in the north woods. There are ghosts of old adobes far out on the *Arizona Highways* desert. There is romance, fantasy and illusion.

The book is not all subtlety and illusion, however. It also comprises realized dreams. In it you will find the detailed story of nine stone, log and earth homes. For each house I've mustered all the facts and traced the entire process from foundation to nailed-down roof. I've also told all the things you ever wanted to know about laying stone, forming stone, stacking logs, ramming earth and laying adobes.

As I worked my way through the three sections that follow, I became convinced that the material I was currently working on was the best. As I wrote about stone, I visualized stone barns, a stone root cellar and a lovely stone studio for my wife. As I got into the log section, I decided that logs were nice, and a rambling, rustic log home in the Alpine style would be perfect. Finally, as I finished up the earth section, I dreamed of building a massive earth house with 3-ft.-thick walls and a large greenhouse facing south. If you asked me now which is the best material to build with, I'd stammer and stutter and tell you that I like them all. If I had to start building tomorrow and I had a good supply of stone, I'd build stone walls. On the other hand, if straight, tall trees covered the land, I would go with logs. With earth, though, I could incorporate the stone and use the trees for beams and rafters. . . . It's no use, I'm still not ready to decide.

It's no small thing to make those decisions involved in buying or building a house. It's the biggest investment that most of us will ever make. It's not easy to come up with the money. It's not easy to find the right place. It's hard to decide on a plan and choose the materials. It's hard to know whether to buy or build. None of the projects described in the following pages was easy. There is a lot of romantic nonsense about buying a hunk of land and hewing your home out of the wilderness written by people with incredible good luck or awfully short memories. I've talked to people who have had good experiences building their dream houses, but I've also heard from those who found that everything cost more, took longer and was harder on their family life than anything they were led to expect by the build-your-own-house books that made it sound so easy. So if you are reading this book as you get ready to build your dream house, remember that there is no easy way to construct a house and that they all have their hidden expenses. Fortunately, time does pass, and it is part of our nature to turn the trying moments into ''adventures'' and funny stories.

I think you'll like the houses I've chosen here. They are a bit out of the ordinary, but they don't look odd. Some of them were built for little money, but they don't look cheap. All are as solid as the stone, logs and earth they were fashioned from, and they're all lived in by their original owners. They were all constructed with strict adherence to the local building codes; some were even built with bank loans and most are insured (though I have heard of some adjusters who initially had trouble putting value on an earth wall).

The story behind the story of the houses in these pages began in the late 1970s when I quit my teaching job and left old friends and familiar territory behind for the adventure of life in the San Juan Islands of northern Puget Sound. My wife and I bought an old waterfront home, enrolled the kids in school and tried to figure out how to make a living. A new friend took me on and taught me about remodeling, and I took pictures on the side for several newspapers. Steve Kenady, founder of the Real Log Cabin Company, hired me to take some photos of his handcrafted log houses for a new brochure. We flew all over the islands shooting the photos, landing precariously in barely level fields, with me hanging out of the window for the aerials. I made some

The author's rammed-earth home, which was built by David Easton in the Sierra Nevada foothills, is shown above. Construction on the Ahwahnee Hotel at Yosemite National Park in California, right, began in June 1926 and, through the labor of 245 workers, was completed in July 1927—a testimony to the timeless beauty and elegance of stonework. The contemporary-looking log house shown below was built by hand-hewn log builder Steve Kenady on Sinclair Island in Washington's Puget Sound.

prints to show my friends what a real log house looked like, and felt the first stirrings of desire for an elemental home of my own.

After several years of island life, we found our sleepy little village waking up and becoming citified, so we began the search for another undiscovered spot. We returned to California to inspect our friend David Easton's first rammed-earth house. Easton needed someone to help him in his fledgling rammed-earth construction company, and he wanted to write a book about his process. He was also having a hard time selling that first earth house. I thought we could make a good book together, and I dragged the family with me for a life in the tall pines of the Sierra Nevadas. We bought the house, Easton and I became partners, and I traded in my framing hammer for a tamper and a front-loading Kubota tractor. We eventually got around to writing the book and some articles, and I discovered that it was much more fun to write about construction than to get all dirty and sweaty doing it. I learned to work on the Easton jobs with a tamper in one hand and a camera in the other, and it was during this period that I recorded the construction process of the Sturgeons' house presented in Chapter 10.

With a couple of articles and a book behind me, I started thinking of myself as an author. I began looking for some houses to write about, crafted buildings that reflected imagination and respect for the environment. I bumped into watercolor artist John Marsh and asked if I might write about his stone house (Chapter 2). It was nicely crafted and comfortable, and John and his wife, Kathryn, had built it all themselves. It seemed a natural for an article in *Fine Homebuilding* magazine in 1983. It was just about this time that I began to think of houses like Kenady's and Easton's and Marsh's as being elemental. I had always been thrilled by the natural grandeur of Yosemite's stone Ahwahnee Hotel and the way it fit into the beauty of the granite valley walls and majestic oaks. It seemed somehow fitting that John Marsh's house would have been designed by a man who was taught by an architect who had been intimately involved with the Ahwahnee project. I was also fascinated by the stone mansion called Vikingsholm (p. 11) at nearby Lake Tahoe, and returned again and again to walk its lovely grounds and to soak up its special magic. What was it about stone and heavy timbers? What was the attraction of earth walls and sod roofs? Of course these are solid structures; they make you feel safe and secure just being near them. There is a sense of permanence and quality here that is absent in most houses built with conventional methods and materials.

With the article about John Marsh's house done and published, I continued to think about the elemental quality that natural, site-specific building materials seemed to possess. I traveled to the Southwest and saw more examples of earth building. I contacted earth builders as far away as Australia. They all spoke of their houses in the same way—that these homes felt solid, looked like they grew there, and were going to be around for a long time. There was a comfort and security that went beyond the structural qualities of the materials. I returned to the Northwest for a closer look at Kenady's log houses and discovered some other log builders. I was surprised that log-house owners used the same fond vocabulary when describing their homes—these houses were warm in the winter and cool in the summer, and they felt safe; the massive walls absorbed sound and so the houses were quiet. John and Kathryn Marsh felt the same way about stone.

There is no way of getting around it. Houses with massive, thick walls look terrific. They're satisfying to build, too—when you lay up the stone for a stone wall, the wall is done. When you ram the earth or lay up the logs, that's it. No sheathing, insulation, sheetrock, paint. It's the concept that form follows function. What you see is what is there. The structural properties of the wall are also its aesthetic. That's not to say that you can't put on some plaster or sheetrock if you want to; the point is that if you like the look of the natural materials, you can leave them that way.

"Elemental" is the word that best describes what these natural materials are to me. They are created by the elements—earth, fire, air and water. When left exposed in a wall, to me they are more attractive than the materials we see in our subdivisions. Their solidity gives them roots.

One of the first stone houses I remember seeing was Jack London's Wolf House in the Valley of the Moon near our first home in California. It was impressive and imposing even as a ruin. (Fire had gutted the house before London moved in.) The early missions every California school kid learns about are still solid and monumental testimonies to the durability of earth. I can still remember tromping around the mountains as a youngster, hiking and panning for gold, and the cabins we found in the woods with fallen roofs and thick log walls. Oh, how we dreamed of taking off right then to live off the fat of the land, to cut the saplings for a new roof, to re-hinge the door and hang some hides and traps up on the logs. These early impressions and romantic yearnings have all become part of my vision. My fondness for the elemental is hopelessly anchored in dreams, and if you look closely as you read about the details of the houses in the following pages, you will see some dimly lit phantasms barely concealed between the lines.

Magnus Berglund
Blue Mountain, California
September 1986

An Introduction to Stone Houses

CHAPTER 1

t's hard to remember the days when there were few shelter magazines, when the only do-it-yourself homebuilding books around were Ken Kern's *Owner-Built Home,* Rex Roberts' *Your Engineered House,* and Helen and Scott Nearing's *Living the Good Life.* Kern covered everything from bio-privies to rammed earth and, as a mason, he had a good chapter on stone masonry. Roberts challenged our beliefs in the elements of building we took for granted—he ridiculed the dug basement, painted wood and windows that could be opened and closed. But it was the Nearings who really got people excited about trying a new adventure, and who made building with stone sound like a viable alternative to stick construction.

"We chose stone for several reasons," the Nearings wrote. "Stone buildings seem a natural outcropping of the earth. They blend into the landscape and are a part of it. We like the varied color and character of the stones, which are lying around unused on most New England farms. Stone houses are poised, dignified and solid—sturdy in appearance and in fact, standing as they do for generations. They are cheaper to maintain, needing no paint, little or no upkeep or repair. They will not burn. They are cooler in summer and warmer in winter. If, combined with all these advantages, we could build them economically, we were convinced that stone was the right material for our needs."

The Nearings did figure out an economical system—poured stone (see p. 13). They went on to build their fencing, outbuildings and house from stone. And when most people their age were moving out of their homes and into retirement communities, these two started over again with another stone homestead, and then amazingly did it a third time when Scott Nearing was in his nineties. Everyone I've

ever talked to about building with stone has read the Nearings' book, even if they have not adopted their methods—the Nearings gave their readers the confidence to get started and the faith to continue.

I admit that I, too, like stone. I live in a rammed-earth house now, but I still like stone, and if I were ever to start over I would seriously consider building a stone house. There is something especially elemental to me in stone buildings. I think of all the stone huts I've visited above the timberline in the mountains—stone and bare peaks are firmly linked in my imagination. In a howling wind, stone walls are as secure as a cave in the mountainside. Even in this age of architectural styles that can only be labeled as 20th-century ostentatious, most people respond to the permanence and stability of a nicely crafted stone house.

By far the most incredible house I visited in the course of writing this book was built of stone in 1529 by Hernando Cortez in Cuernavaca, Mexico (photo, p. 6). More of a palace than a house, it is rich in Mexican history—Zapata was there in addition to Cortez, and there are the marks of revolution and executions and the legend of buried treasure somewhere on the property. Now owned by Jorge Fenton of Mexico City, it is one of the oldest buildings still in use in this hemisphere. An aqueduct of stone dates back to the time the building was used as a convent and church, and there are acres of stone outbuildings, which Fenton is currently restoring. Everywhere in the cool interior of the house can be seen the permanence of stone. Swimming pools and automobiles on the cobblestone lanes have brought this place into the 20th century, but the classic arches and monolithic walls speak books about the enduring qualities of this natural building material.

Styles of stonework

Most people who experiment with laying stone find it to be an addictive and sometimes even a magical building process. Obviously, laying stone also involves sweat, sore muscles and bruised fingers. Lifting, moving and setting 50-lb. stones is not for sissies. Hauling sacks of cement, shoveling sand,

The D.L. James house sits 80 ft. above the sea on the California coast, exposed to battering storms. Architect Charles Greene recessed the arched windows to shelter them from the elements. The house has no man-made foundation—the walls appear to grow from a trench the mason hollowed in the bedrock. Photo by Charles Miller.

Stone arches, circular and oval windows, pillars, and thick walls mark the work of the Spanish and Indian masons who labored under Hernando Cortez to build the walls of the Cuernavaca Estate in Mexico in the 16th century. Shown here is the main entry to the house.

handling and cutting hundreds of feet of reinforcing steel is all physically demanding work. The magical part of the process sneaks up on you. Sometimes it is as startling as finding out that all the stones in the last load fit together as if each were a part of some giant jigsaw puzzle. Sometimes it is more subtle. You realize that as your eye records an odd-shaped space, your hand automatically responds by picking up the right stone. Gradually you realize that you chose the rocks in your stockpile way before you had any idea of the spaces they would fit. That is the point at which it is possible to take yourself out of the way and simply lay stone.

When I look at a stone wall made by a beginning mason, I like to think that I can see where the magic began. I can certainly see it in my own work. At the bottom the stones seem to be looking for a sense of pattern. The jigsaw-puzzle look of good work isn't there. Some stones obviously should have been saved for a better spot. Mortar oozes from the joints. There are spots where the mortar is scraped evenly all around a stone and spots where it has been neglected. The beginning mason is not sure how to handle the mortar. How do you trim the excess? Should you try to fill all the voids between the stones? What looks best? When is the best time to dress up all the joints? Masons call this final treatment of the mortar lines ''pointing,'' and it is a key step in creating a professional-looking stone wall (see p. 10).

It usually takes only a couple of feet of wall for the beginning stone mason to get into the rhythm and magic of laying stone. All of a sudden there are handsome stones in pleasing patterns, an attractive balance of large and small stones, and mortar that is pointed consistently—you can tell where the mason started to enjoy the work.

As you become aware of stonework, you will see that there are a variety of styles possible. There are walls built of cut and shaped stones that are orderly and symmetrical. There are walls of round, river-washed cobblestones. Some walls, such as those in Vikingsholm, shown on p. 11, are built with huge boulders. Others, like John and Kathryn Marsh's house (which is discussed in Chapter 2), have a variety of sizes and shapes. Much of the old stonework in the mining towns of the California foothills is done with flat rock in neat layers, and looks almost like brickwork.

Some masons plan their patterns and work carefully to repeat a shape or sequence as they go. Others are more intuitive, placing the stones they have gathered in relationships of pattern, texture and color that develop as the wall rises. The nature of the stone determines the pattern to some extent. Square or rectangular stones lend themselves to geometric patterns, whereas random sizes and shapes lead to flowing, organic patterns. Some masons lay the stones so that each course adds a straight horizontal line of mortar as in a brick wall. Others favor an uncoursed look.

The style you choose is a result of your preferences, the type of rock you have to work with, the time you wish to spend and the level of your skill. Random, uncoursed stones of varying sizes laid in a freeform pattern is probably the easiest style for the uninitiated. A mortar line that is pointed to the surface of the stones or slightly recessed is also fairly easy to handle. As you concentrate on putting a nice surface to the face of the wall and a more or less flat side up and a flat side down, the patterns start to take care of themselves. As you work, you begin to develop a feel for placing large and small stones, rough ones and smooth ones, and you start to notice the losers before you even lay them in the wall.

The houses in the following three chapters were the first stone projects of each of their builders. Three different methods of laying up the walls were used, but all the builders used found stones in random, uncoursed patterns. John Marsh (Chapter 2) pointed his mortar so that the rounded faces of the stones stand out in relief. The Brabons' house (Chapter 3), even though it was built of formed, poured stone, is pointed very much like Marsh's house, although the Brabons waited until they actually moved into their house to do it. The Monsons (Chapter 4) avoided the whole issue of pointing by not filling in the gaps between stones— the walls are rugged-looking and resemble the dry-laid stone walls that line the Sierra Nevada foothills.

This stone house, located on the outskirts of Mexico City, was designed by Ignacio Colin and built with the help of Mexican stone masons. Many of the timbers and much of the wood trim in this house were recycled from Indian plank houses. The stonework is a good example of the visual variety inherent in an uncoursed pattern.

This wall was built of cut and shaped stone laid in an orderly, coursed pattern.

John and Kathryn Marsh's house looks like a natural outcropping of its hillside site on Bald Mountain in California. It was the Marshes' first stone-laying experience, and they spent over four years planning and building (see Chapter 2). Photo by John Marsh.

The stone in this building is laid in flat, neat courses and resembles brickwork.

Laying the stone

A good way to learn about stone masonry is to look at some projects in your area. Talk to the masons about which stones to use and which not to use, and which mortar mixes they favor. You'll probably discover that freshly quarried stone is more fragile than old stone, which has been exposed to the air for a long time and has hardened. You can also learn plenty just by observing the stone on your building site. When it is plentiful, fieldstone or river rock generally presents a wide variety of sizes, colors and shapes. Here in our area there is a lot of volcanic rock that the natives call rhyolite. It looks very hard, but in fact it is highly porous—it soaks up water easily, then cracks and bursts in a freeze. The rock is plentiful, colorful and deceiving. We used big hunks of it to border our driveway and walkways in our first years here, and each winter more of it breaks down into little pieces. Highly stratified or cracked rock will behave the same way in a freeze, so choose your stone carefully.

Mortar is what glues the stones together into a wall, and portland cement is the key ingredient. The most common mix is one part cement to three parts sand. Some builders add a small amount of fireclay to give the mortar a buttery consistency and to make it stick better to the stone—I usually add one-half part clay to the above formula. Whether you use clay is really a matter of personal preference, as it affects the feel of the mortar more than its strength.

A wheelbarrow and a hoe are all you need for mixing. When putting your mortar recipe together, mix the dry ingredients thoroughly before you add the water. This step is important, as it actually makes the mortar stronger. The amount of water you use is critical, and depends a lot on how wet the sand is to begin with. It's easy to add too much water—one moment the mix is too dry, and then just a squirt of water later it is way too wet. It usually takes only a couple of runny batches before you get a feel for adding the water—when you reach the critical stage, remember that just a couple of drops more will make the mix right.

If the mortar is too dry, it will not spread when you put it on the wall nor will it stick to the stones. If it is too wet, it will run down the faces of the stones and will not support their weight. In between these extremes is mortar that is creamy enough to work and stick to the stones, and stiff enough to stand without dripping over the completed work.

If you've ever worked with concrete, you will have no trouble mixing and taking care of mortar, as the cement in mortar behaves the same as it does in concrete. A freeze will spoil uncured mortar, just as it will a concrete foundation wall. A long, slow cure is best, which means a cool, damp environment for the mortar. In hot weather keep the mortar damp by covering it or by spraying it with water, but bear in mind that a heavy spray will wash out the mortar or, at the very least, leach out the cement.

To lay up a stone wall you'll need a wheelbarrow full of mortar and a stockpile of stones. But before you begin work, you should realize that even a small garden wall is going to take a lot more stone than you ever suspected. We have a joke here on our little farm about going after another ''last'' load of rock. When we built the stone corners on our rammed-earth barn, we made at least seven ''last'' runs down to the river with the pickup truck. It takes about 30 tons of stone to make one 10-ft.-long by 10-ft.-high by 2-ft.-thick wall.

You'll also need a couple of simple tools. A broad, pointed mason's trowel is good for digging mortar out of the wheelbarrow and placing it on the stone. I've also found that a small pointing trowel is helpful for slapping small loads of mortar into unfilled openings. A pair of heavy, rubber-coated gloves will keep the skin on your fingers and prevent it from drying out.

The actual laying of the stone is fairly straightforward. Most masons find it helpful to test-fit the stones before setting them in the mortar. That way it's possible to experiment with different placements of a stone. Once you've got the stone you want and know which way it fits best, put down a trowelful of mortar in which to bed the stone. Don't use too much mortar, or it will ooze out and fall to the ground; but don't use too little either, or the voids won't fill and the stone will sit on the stone beneath it rather than on a bed of mortar. And don't spread the mortar out—the weight of the stone will push the mortar around itself and the stone below. Now put the stone in place. Don't try to move it once it's laid, because the mortar makes a perfect suction fit and if you move the stone you'll weaken the bond. (If you have to move a stone, pull it off the wall, remortar and reapply the stone.) Once the stone is in place, scrape away the excess mortar with a trowel, and use it to fill the vertical joints or any voids. If any mortar drips on the stones below, don't try to wipe it up right away—it will smear all over. Leave it until you are ready to point, and you'll find that it will flake right off the wall without leaving much of a mark.

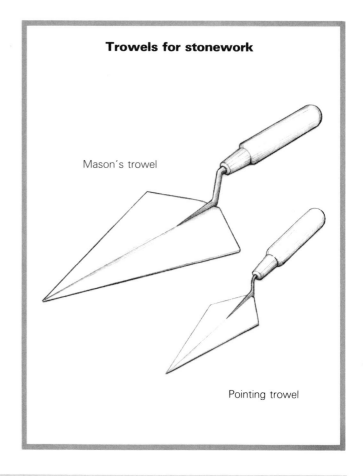

Trowels for stonework

Mason's trowel

Pointing trowel

The front and side walls of the Butte Store, Butte City, California, built in 1857, are all that remains of this gold-mining town. The angle iron at the top is an attempt to hold the old walls together (above). Notice how the front wall is pulling away from the side wall (right). Other cracks seem to be developing where the "one over two, two over one" rule was not followed.

Modern-day stone masons have a real advantage over their forefathers. The cement-based mortar that's available today is generally much stronger than the old lime-based mortars and it is certainly stronger than the mud mortar that was often used 100 years ago. However, one of the first rules of good stone-laying is still to lay the stone as if you expected the mortar to eventually fail, that is, to assemble the wall in such a way that gravity alone will keep the stones tightly stacked even if the mortar should turn to sand. For example, don't stick odd-shaped stones onto the face of the wall with only the glue of the mortar for support. And don't lean heavy stones outward. Either seat them so they lean inward or so their weight presses straight down on the wall.

There's another rule of thumb to keep in mind when laying stone: One over two, two over one. This simply means that you should always bridge the gap between two stones with another stone, so that an unobstructed vertical line of mortar does not travel up the wall. If you don't do this, there's a good chance the wall will crack along the vertical line. It's surprising to me how many masons have either never heard of this rule or have chosen to disregard it. It's also surprising that a lot of this "bad" work has not cracked, a tribute more to the use of reinforcing steel and superior modern cement mixtures than to any magical powers of the masons in question.

A stone wall laid with good mortar, carefully stacked stone and some judiciously placed ⅜-in.-dia. rebar is going to stand for a long, long time. The stone walls built by the masons who followed the gold rush to California in the 1850s are good examples of gravity doing its work—even with crumbling mortar and minimal reinforcement, the walls still stand.

The "one over two" rule

In an incorrectly laid wall, mortar is allowed to travel in an uncrossed vertical line, along which the wall will likely crack.

For the strongest stone wall, follow the rule "one over two, two over one" when placing the stones.

Pointing As the mortar begins to cure, it gets grainy before it gets completely hard. It's at the grainy stage that pointing is most easily done. There are different styles of pointing, and each creates a different effect. The mortar can be brought right up to the face of the stones, or it can be scraped back into the joints so the stones stand out in relief—the first approach gives a flat, clean-looking surface, while the latter creates nice shadow patterns that outline all the stones. In buildings making a formal architectural statement, a mortar joint that is built up to extend beyond the surface of the stones is often found. This type of pointing requires special tools and cement-rich mortar. It is not for the beginner—besides being more work, it also draws attention to itself and thus is best left to an experienced hand.

When you're ready to start pointing, the tools you'll need are a wire brush and a thin pointing trowel. I use the trowel for scraping and digging out mortar from the joints, and the brush for cleaning off the stone faces and combing the mortar to make a textured joint. (Ken Kern recommends the stiff-bristled brush sold for cleaning auto parts for smoothing the joints.) In warm weather the mortar will be ready for pointing in a few hours. In cooler weather you may be able to wait overnight, but don't wait too long, or you'll need a hammer and chisel to chip away the hardened mortar. If you favor really clean stones, wash them with a dilute solution of muriatic acid. The acid dissolves lime (don't use it on limestone) and releases cement from the stone. Hose it off when it has stopped fizzing and keep it out of your eyes and off your clothes. None of the builders I talked to used the acid treatment except Bob Monson (Chapter 4), who used it to clean up his pebble floors and countertops.

Pointing mortar joints

Depending on the effect desired and the mason's pointing skill, the mortar can be recessed to various depths, brought up flush to the surface, or built up to extend beyond the face of the wall.

The time factor

It's best when you start a stone building to realize that it is going to take longer and require more patience than any single project you've ever undertaken, aside from raising your children or writing an encyclopedia. Building a stone house cannot be done in rabbit time, the way stick houses in suburban subdivisions are built by people who know that time is money. Rather, stone houses are built in elephant time. The person who builds one is definitely not in a hurry. He or she may be a rabbit-timer at the start, but will be either crazy or an elephant-timer at the finish. Stones have to be found to fit in their special places. Mortar must be mixed, and you always seem to run out just when you're on a roll. Then the mortar takes time to cure. The experience of building with stone is further enhanced by the delays that affect all builders—by the trucks that can't get down your road, the materials that are promised but not delivered, the electrician who doesn't show up the day you want to do the wiring.

There is one way around the time problem in stone masonry, and that is to do what Mrs. Lola Knight did in 1928 on the shores of Lake Tahoe, California. Mrs. Knight had decided that she wanted a stone-and-timber building in the Sierras for a summer hideaway and that she wanted it built like a Viking castle. Being a woman of considerable means (and a rabbit-timer), she hired a Swedish architect and 200 craftsmen to build Vikingsholm, which you can see at the top of the facing page. They did it in one summer. They laid up gigantic granite boulders, shaped and hand-carved the timbers, hand-painted the walls and beams with Scandinavian motifs, and created a fanciful and inspiring piece of elemental architecture.

Not all builders with the financial means are so impatient. The house that D.L. James built on the California coast in 1918 took several years to complete. Designed and supervised by Charles Greene, who with his brother, Henry, had done much to define the Arts and Crafts style of American architecture, the house literally grew from the bedrock of its spectacular site (bottom photo, facing page). James expected it to take two years, but his architect seemed to be making the project a lifetime commitment. After five years, James was finally able to enjoy his house, even though everything in the original drawings had not been built. Of course, the time it takes to make a building is not critical to the results. It's just that it is hard to hurry the traditional stone-laying process. Vikingsholm, more than 50 years later, looks no more like a rush job than the James house does.

Vikingsholm at Emerald Bay on Lake Tahoe, California, is a blend of huge granite rocks and heavy timbers (top left, facing page). Its morning room (top right, facing page) faces the beach. The original mortar contrasted nicely with the dark granite boulders, but recent refurbishing by the state, which now owns the estate, has covered the old work.

The massive James house (bottom, facing page) was one of architect Charles Greene's most ambitious projects. Over a period of several years he supervised its construction from the golden-hued granite found on the site. (Another view of the house is shown on p. 4.) Photo by Charles Miller.

Using the poured-stone construction method he invented, architect Ernest Flagg created a series of small cottages on his estate in Staten Island, New York. Bow-Cot, above, was so named because it curves with the road it watches over. Photo by Andrew Popper ©1981.

There is, however, a type of construction that can speed the stone-building process without requiring a large infusion of money and an army of workers. This is the system of poured stone made popular in the 1920s by New York architect Ernest Flagg. (The house in Chapter 3 is made of poured stone.) Flagg, an early proponent of appropriate technology and affordable housing, felt that because stone was so readily available and such a sturdy material, it should be used to construct inexpensive housing. In other parts of the world stone was used by simple people for simple housing, but in America, cut and quarried stone was expensive and was laid slowly by highly paid masons. Flagg wanted a method that even inexperienced builders working on a shoe-string budget could use, so he developed a system for poured stone. The form was basically a big box built from parallel boards and braced uprights. Flagg placed the stones with their nice sides out against the outside boards and then shoveled concrete in behind them and up against the inside boards to make a smooth concrete interior wall. The Flagg house developed into a recognizable style: The houses were low (no heavy lifting above the builder's head), cellar space was minimal (no heavy digging), and a slab floor was poured where possible (a one-step process for the beginning builder). The houses also had door and window frames built into the walls (no fancy finish-carpentry skills required). The low walls easily supported a gable roof without trusses or wall ties. For the corners of his houses, Flagg stacked cast-concrete corner blocks having large vertical holes, then fit reinforcing rod through the holes to hold the corners firmly together.

The Flagg method convinced the Nearings that stone could be economically and efficiently used to construct their buildings. They used a batch of 18-in.-high forms of various lengths, and improved on Flagg's method by making their forms lighter and easier to move. They placed the stones in the forms, poured in the concrete and worked it around the stones. The forms were stacked atop each other—as the bottom course cured, its form was freed and placed on top of the second form. The forms were light enough that one person could handle a 14-ft. length, and with a lot of forms on hand, the Nearings could set up around a building and leapfrog to the top of the wall in no time.

Karl and Sue Schwenke of Wells River, Vermont, successfully adapted the Nearings' methods to their own needs and built a lovely, 1700-sq.-ft. stone house. They poured 24 to 35 running feet of 19-in.-high wall a day using a basic wooden slip form that was 8 ft. long and 19 in. high. Their walls range from 14 in. thick at the base to 10 in. thick at the top. For horizontal reinforcing, the Schwenkes ran discarded steel sawblades from a nearby rock quarry on 18-in. centers. For extra strength at the corners, 8-ft. lengths of double-pointed barbed wire were run in between the sawblades. The Schwenkes wrote a book about their experience, called *Build Your Own Stone House*.

Poured-stone walls as described here contain a lot of concrete, often as much as one-half to two-thirds the volume of the wall. More rock and less concrete can be used by dumping broken rock and pieces of hardened concrete into the form. This is often called rubble fill, though some writers use the word ''rubble'' to describe the fieldstone or random rocks used to build houses such as those discussed in the next three chapters.

The interior of the walls in a poured-stone house must be finished, as the raw concrete does not make a very attractive surface. Flagg and some other early experimenters simply plastered the interiors and painted or wallpapered them. More recent builders have tried various methods of insulat-

Using a basic wooden slip form, Karl and Sue Schwenke built this 1700-sq.-ft., poured-stone house in Wells River, Vermont. Photo by Karl Schwenke.

Corner stones that lap into the adjacent wall add strength and earthquake-resistance to a stone structure.

ing and finishing the concrete walls. The Nearings placed vertical strips of wood (furring strips) right in the concrete as they formed the walls. The strips were flush with the surface of the wall and allowed for the nailing on of horizontal strips of wood. The Nearings put building paper over the horizontal strips and nailed wood paneling on top, creating dead air-space for insulation. Other builders have used insulation board of one kind or another directly bonded to the concrete in the forms. The Schwenkes recommend another method for cold climates. They built a 2x4 stud wall inside the stone walls with a ½-in. airspace and attached foil-faced insulation between the 2x4s. They then finished the wall with sheet-rock and cherry paneling. The Schwenkes also talk about neighbors in Vermont who have used a variation of the insulation system proposed by Rex Roberts in *Your Engineered House*. They used 2x4 furring strips set in the concrete for nailers and then nailed two separate layers of horizontal 1x3 strips on these with layers of foil separating them. This creates two dead airspaces, separated by the foil. A finished wall of sheetrock or paneling is nailed on last. The Brabons in Chapter 3 used the double-wall system recommended by the Schwenkes.

Requirements of the stone house

The engineering problems associated with stonework are generally the same as those for concrete, brick, adobe or rammed earth. In the codes and textbooks, these materials are usually lumped together under the heading of masonry construction, and they share common characteristics. Walls made from these materials generally have high compressive strength (they will support a great deal of weight), but low tensile strength (they don't bend). This lack of flexibility can be a real problem in earthquake-prone areas such as California. In such areas, masonry builders are required to put steel reinforcing into their walls. All of the stone houses discussed in the following three chapters were built in earthquake country and all use a system of tied horizontal and vertical rebar.

The mortar itself is an important element in a stone wall. If it does not lock the individual elements together, then during a tremor the wall will crumble. Hence it is not enough to stack the stones two over one and place them carefully so that gravity holds them in place. They must also be firmly locked together with a good mortar that has been properly prepared, applied and cured.

Stiff masonry walls that are locked together at the corners make what engineers call a moment-resistant frame. All the walls act as one unit, so in an earthquake the walls theoretically move as a large box rather than as individual walls. But if the walls are not tied together at the corners, they can fall out or in. Some stone masons therefore use corner stones that overlap into the adjacent walls. Some put metal ties between the stonework that runs around the corners. Builders who pour stone form their corners in one session to avoid cold joints (places where you stop pouring and the concrete hardens before the next pour) right at the corner. The builders in the next three chapters depend on their horizontal bands of reinforcing steel to lock the corners to the rest of the walls and to make the building one solid unit.

Another consideration in building masonry walls is the size and frequency of openings. Large openings could interfere with the ability of the structure to operate as a moment-resistant frame. Concrete headers with extra-heavy rebar tied into the rest of the reinforcement are critical. In a house with complex shapes and lots of corners and openings, the services of an architect or structural engineer are called for.

Obviously, masonry walls are much heavier than conventional house walls and so require stronger, larger footings. The stone builder must calculate the weights of the walls (as well as of the roof and the rest of the house) and design an appropriate footing. Local soil conditions and frost depths must also be considered. Building departments in many areas require an engineered foundation system for stone or earth houses simply because they don't have the experience to evaluate them.

Because of the problem of lifting heavy materials, it is easier to build low stone walls than high, two-story ones. High walls can be built, but they create more complex structural problems. They may need cross ties to hold them safely erect. The lower part of the wall will have to be thicker to support the weight of the second story. Generally, a 12-in.-thick wall is sufficient for one story. A two-story wall should be at least 16 in. thick at the bottom and 12 in. on the second story. A general rule of thumb is that wall height for a single-story building should not exceed a ratio of 10:1 to wall thickness, that is, a 12-in.-thick wall should not be more than 120 in., or 10 ft., high. Be aware also that different roof systems put different pressures on walls. For example, a gable roof pushes outward at the wall tops and requires some sort of tie or roof truss for lateral support. A stone house designed as a small, short rectangle will support a roof with the strength of its wall mass and corners, while a structure with long walls and many window and door openings is going to need some extra support and engineering. A thicker wall can be more or less self-supporting; a thinner wall will need help in the form of buttresses or ties. As you read the following chapters, notice how these particular problems of stone construction have been solved by the different builders.

A question always asked of people who build with stone, logs or earth is: Where does the wiring and plumbing go? For all three types of building, the answer is pretty much the same. Wiring and pipes should ideally be put in the floors and/or divider walls. (Plan carefully, though, because if you decide you need extra wiring and plumbing after the house is up, it will be impossible to get into the monolithic walls.) Outlets and electrical boxes can be mortared into stone walls and the wiring run through embedded conduit. Plumbing should be left where it is accessible in case of leaks— under cabinets and in wood-frame divider walls; if there is no alternative route, copper or plastic pipe can be embedded in a stone wall. In the places where a hole is needed through the stone wall for a drain pipe or electrical service, a short piece of plastic pipe can be laid across the wall's width during construction and mortared in place. Again, think these places out beforehand, for there is no way you are going to come back later and drill a 5-in.-dia. hole through a stone wall for a dryer vent.

Door and window openings must also be carefully planned. If a door is too big, a log wall can be cut out and enlarged, an earth wall can be chipped out, but a thick stone wall is not going to be altered without tremendous effort. It's a good idea to buy or make your windows or doors before you start the stone walls and to lay or pour the walls against their framework. Where wood framing is attached to the stone, drive spikes through the wood, then work the mortar around the spikes. Even if the wood shrinks over time, it will remain firmly bonded to the stone. It's also advisable to treat the wood that is in contact with masonry or to use a decay-resistant wood like redwood. Masonry/wood connections are places where moisture is likely to accumulate.

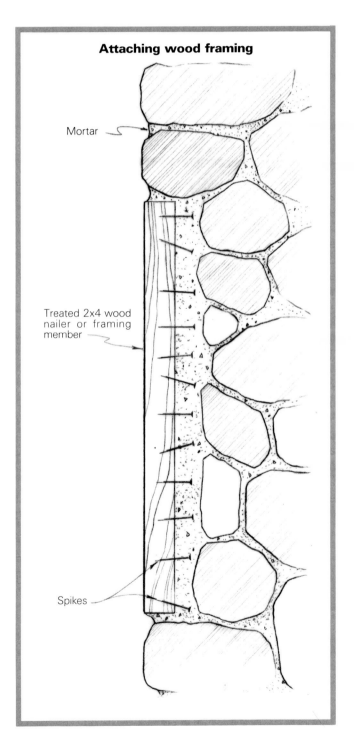

Attaching wood framing

Mortar

Treated 2x4 wood nailer or framing member

Spikes

This house, designed and built by Ignacio Colin, is made of volcanic rock and is plastered inside and out with cement plaster. It is located in an exclusive subdivision on the slopes of Mt. Ajusco in Mexico City.

The Sierra Club's La Conte Building in Yosemite National Park, California, is made from cut granite that echoes the stones in the high cliffs of the valley's rim.

The stone houses in the following pages are not something you'll find in an American subdivision. These are the romantic alternatives for those of us who look to the values of a previous era. Only the superficial aspects of these styles reach out to the mass market. For the fashion-conscious, there is milled siding that imitates logs and the ever-popular prefabricated milled-log house. There are the red-tiled, brown-stuccoed subdivisions that are designed to look like their adobe cousins. Perma-stone and other variations of the glued-on-rock look go through spurts of popularity. By adding these stylistic elements to our houses, we seek the attendant illusion of quality, permanence and beauty, and somehow end up with the opposite effect. There is no way to get the strength and character of a grandfather tree by gluing the bark to the outside of your house, or the romance of stone by mortaring up a veneer of rock. The pleasures and the spirit of the materials are reserved for people like the homeowners in the following pages who have actually built with the material.

The curving north and west walls of this house are 2 ft. thick and 22 ft. high at the peak. Built by Rick Hayton and Sebastian Eggert on the Olympic Peninsula in Washington State, the massive structure required 200 tons of irregular stone. Photo by Charles Miller.

The House on Lupine Hill

CHAPTER 2

When John Marsh left his San Francisco design firm to pursue his life as a watercolorist in the quiet solitude of California's Sierra Nevada mountains, he and his wife, Kathryn, bought 20 acres on the top of Bald Mountain with a beautiful view of the Mokelumne River, the 6,000-ft. Blue Mountain and the snow-capped peaks of the Sierra Crest. In the spring the site was covered with blue mountain lupine, green grass, live oak and red-trunked manzanita. And rocks—you couldn't walk two feet in any direction without tripping on a big hunk of granite. Products of an ancient riverbed, the rocks were mostly rounded, and after years of exposure to the elements, they were beautifully colored with lichens and mineral deposits.

The Marshes decided to use the stone on the site to build their house for a number of reasons. Construction costs would be cheap—the rocks were there for the taking and cement was a dollar a sack at the time. Even though they had no building experience and knew the labor involved in stone-building would be intensive, they felt that the skills could be easily mastered. In addition, John had grown up in New York's Adirondack Mountains and rural Connecticut, where fine old stone houses abound, and he wanted to create some of the feeling of those places in his own home. To Kathryn, stone was especially appealing because of its durability—her childhood home had burned to the ground in a southern California brushfire, and she wanted the security of a house that wouldn't burn.

A stone-lined walkway leads down from the crest of Lupine Hill to the stone home built by John and Kathryn Marsh. Perched at a 4,000-ft. elevation in the foothills of California's Sierra Nevada Mountains, the 900-sq.-ft. house was the couple's first experience with this type of construction. They camped out on the site for four years while working on the project.

It took the Marshes four years to build their home, christened Stonehaven. (Only two of those years—and 300 sacks of cement—were spent in actual stonework.) At the outset, they moved into a temporary campsite on their new property and began gathering stone. They carried it. They wheelbarrowed it. They hauled it in a flat-bed trailer towed behind their jeep. As they got more settled and completed the wood-frame screen home on the mountainside in which they would live until Stonehaven was finished, they were able to spend entire mornings picking and hauling rock along the ridgetop. Soon they had a 60-ft. by 10-ft. pile almost 8 ft. high at the building site (which would last only until the walls were about 4 ft. high).

Talking with local masons, Marsh discovered that his major problem in laying stone would be getting a proper cure in the dry summer heat of interior California. The same arid conditions that make the forests tinder-dry during this season also make mortar dry out much too quickly, and such mortar will likely fail. So Marsh saved scrap cardboard, old gunnysacks and building plastic so that once he began laying stone, he could cover the walls and keep the sun off the freshly laid mortar.

The Marshes' house was designed by their longtime friends, San Francisco architects Zach Stewart and Daniel Osborne. They faced two challenges—coping with the lack of electricity on the building site and keeping the house dry. Generators easily solved the first problem; the second was more difficult. The ideal spot for the house would have been on the ridgetop next to the road and the well. To accommodate a gravity-fed water system, however, they placed the house 15 ft. downhill from the water storage tank. Unfortunately, the same gravity flow that would pressurize the water lines to the house would also force water from the hillside through the stone walls and up through the concrete slab and footings. Surface water could easily be diverted with drain tiles and rock, but Stewart and Osborne were concerned about water seeping under the walls and slab. Their solution was a combination of drains and membrane, which has kept the house bone dry for the past 14 years.

Marsh house—Lupine Hill

Southwest elevation

Southeast elevation

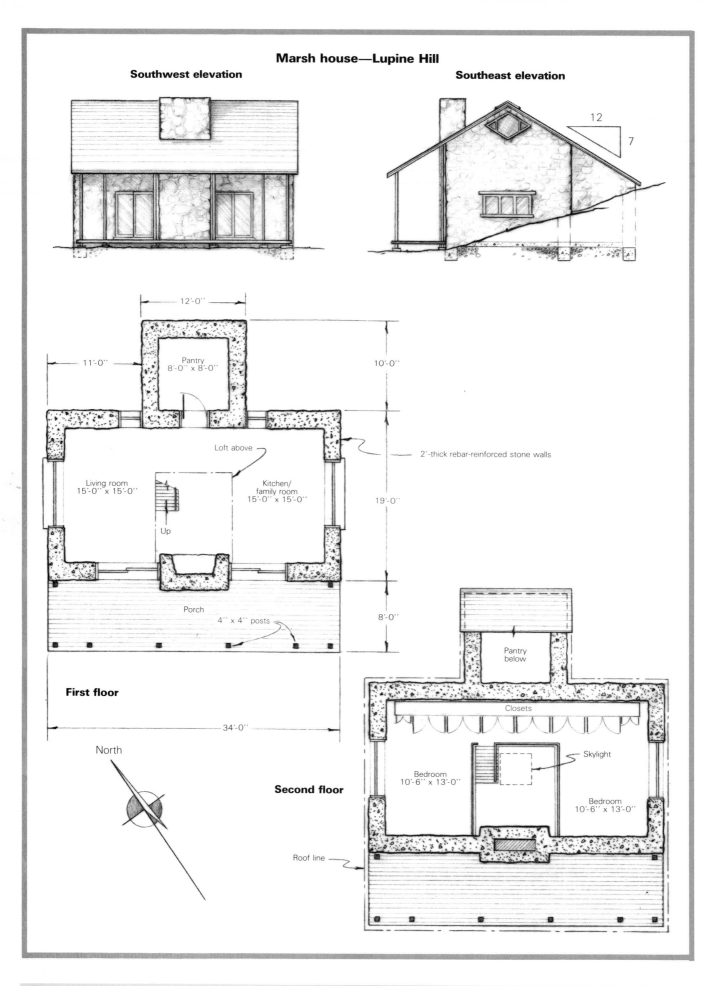

12
7

12'-0''

Pantry
8'-0'' x 8'-0''

11'-0''

10'-0''

Loft above

Living room
15'-0'' x 15'-0''

Kitchen/
family room
15'-0'' x 15'-0''

Up

2'-thick rebar-reinforced stone walls

19'-0''

Porch

4'' x 4'' posts

8'-0''

First floor

34'-0''

North

Second floor

Pantry
below

Closets

Skylight

Bedroom
10'-6'' x 13'-0''

Bedroom
10'-6'' x 13'-0''

Roof line

The east wall (above left) catches the first sun as it rises over the High Sierra. The kitchen window and the diamond window were both designed and built by Marsh. On the north side, the pantry's tile roof nearly meets the hillside (above right).

The railing around the loft bedrooms consists of pegged 4x4 posts, 1x2 edge pickets and a 2x10 cap.

The membrane is a 35-ft. by 50-ft. sheet of 12-mil polyethylene plastic. The entire house—foundation and all—sits inside this giant plastic envelope. At the corners of the building the plastic is folded like a bedsheet, and it extends 6½ ft. up on the back wall and almost 4 ft. up on the side walls. At the base of the footings (outside the envelope) is a standard drain system using perforated plastic drain line and gravel fill.

Stonehaven's 2-ft.-thick stone walls are an impressive 20-ft. tall at the gable ends. The house has 900 sq. ft. of living area, but looks much larger than it is, thanks in large part to an emphasis on things vertical. The house includes two loft bedrooms connected by a walkway, and on ground level a kitchen, a pantry and a large living area. The center of the house is open to the loft area, leaving the exposed stone walls of the gable ends, the leaded-glass casements made by the Marshes and the diamond-shaped windows under the roof peak visible from most of the first floor. The massive stone fireplace rises through the south side of this loft opening. The exposed joists and rafters give the house plenty of interesting overhead detail.

The ridge of the clay-tiled gable roof runs along an east/west axis. The house is cut into the shoulder of Bald Mountain on the north, and the uphill wall jogs to incorporate the 8-ft.-square pantry. The pantry's shed roof is nearly at right angles to the slope of the grade on this side of the house, and comes within a foot of meeting the ground at the overhang of the pantry wall. A full-length porch shades the south side of the house, but winter sun enters through French doors on both sides of the fireplace. The floor of the house is covered with hexagonal clay tiles made by Kathryn Marsh (p. 27). The house is comfortable, dry and easy to maintain. It is cool in the summer and heated with ease by a woodstove in the winter.

The center of the house is open to the loft. John Marsh designed the ladder on the living-room side; Kathryn manufactured 3,000 clay tiles for the flooring.

The foundation

Two things that cause California builders to loose sleep at night are earthquakes and sliding hillsides. Stewart and Osborne had specified an interlocking rebar system to take care of the stresses caused by quaking earth. But the potential slippage of a house that was perched on a steep slope and weighed hundreds of tons was a different matter, and called for a foundation system that would keep the building securely moored to the site. Originally the plan was to build the house on a level pad carved into the hillside; a perimeter foundation dug down to undisturbed soil would loop around the pantry area at the back of the house. The plan was modified, and the system made stronger, when Marsh hit rock during excavation of the pantry area—the 2-ft. by 2-ft. perimeter foundation wound up looping around an 8-ft. by 8-ft. by 2-ft. rock island. Marsh also enlarged his basic 2-ft. by 2-ft. trench at the fireplace, which needed a 4x8 footing. The concrete slab floor (over which the tile was eventually laid) contains reinforcing wire mesh, and was poured on top of layers of pea gravel, sand, plastic membrane and sand.

Marsh elected to dig the trenches by hand because he didn't want footing forms puncturing the plastic membrane, but this meant he had to be careful to keep the trenches neat and square. After the backhoe had scooped out a level building site, Marsh began to dig. He chipped out a few feet of rock a day on the back wall with a 25-lb. bar, an old ax, and a pick and shovel. Digging the side and front walls was easier, since most of this labor was in fill dirt from the backhoe excavation. Because this fill would make an unstable footing base, though, Marsh dug a series of 18-in.-dia. pier holes 3 ft. on center along the front trench. These extended 18 in. down from the bottom of the trench into undisturbed soil.

Once the footings were dug, the Marshes put down pea gravel, covered it with sand and rolled out the membrane from the front of the house to the back. For the pier holes, they made liners from plastic and attached them to the membrane with mastic. When this was done, they rerolled the remainder of the plastic, putting it out of the way until it was time to apply it to the back and side walls. They carefully spread a layer of sand atop the membrane in the area where the slab would be poured.

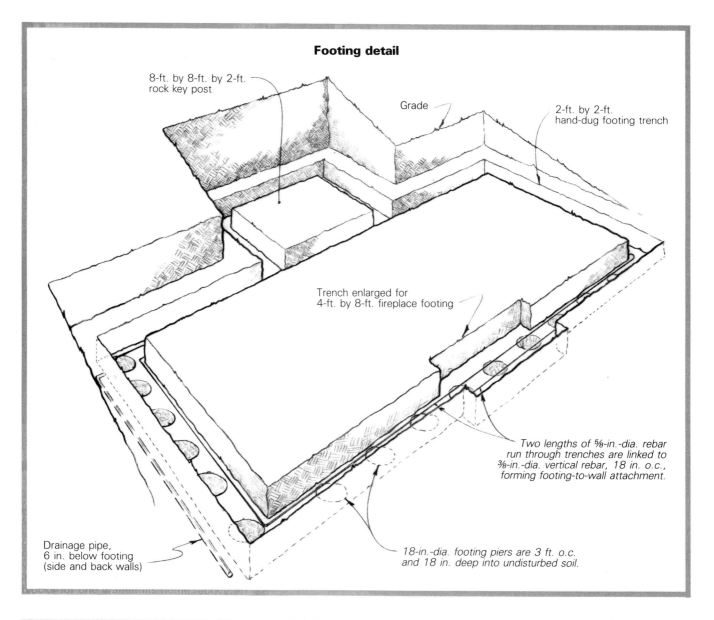

Footing detail

8-ft. by 8-ft. by 2-ft. rock key post

Grade

2-ft. by 2-ft. hand-dug footing trench

Trench enlarged for 4-ft. by 8-ft. fireplace footing

Two lengths of ⅝-in.-dia. rebar run through trenches are linked to ⅜-in.-dia. vertical rebar, 18 in. o.c., forming footing-to-wall attachment.

Drainage pipe, 6 in. below footing (side and back walls)

18-in.-dia. footing piers are 3 ft. o.c. and 18 in. deep into undisturbed soil.

Marsh then laid two runs of ⅝-in.-dia. rebar over the membrane in the footing trenches. To tie the footings to the walls, he tied in vertical pieces of 10-ft.-long #3 (⅜-in.-dia.) rebar to the horizontal bars on 18-in. centers. (Eventually, the vertical bar would run all the way to the top of the walls.)

The footing pour took 15 yd. of concrete. When it had set up, Marsh bent over the protruding tops of the vertical rebar and drew the outside edges of the membrane back over the footings toward the interior of the building and out of the way. Then he dug down 6 in. beneath the footings on the side and back walls, and laid perforated drainpipe in a bed of gravel. The membrane was moved back to the outside of the footings and protected with 2-ft.-wide strips of plywood until the walls could be laid up. Marsh then straightened out the rebar that he had bent.

With the footings poured, Marsh prepared to put down the floor slab. The three doorjamb assemblies were set in place and temporarily braced. (Nails left sticking out of the bottoms of the thresholds locked them into the poured slab.) Two courses of stone were then laid all around the house to contain the slab. More sand was added and leveled, and the 4x4 reinforcing wire was laid down. Many builders who pour a slab and footing system like Marsh's do the whole process in one monolithic pour. Others pour the slab separately and make sure that it doesn't bond to the foundation or stem walls, so that if the walls settle or move, they won't push down on the slab and crack it. By mistake Marsh bonded the slab to the stone stem walls—his architects had wanted him to isolate the slab with an expansion joint, but he missed it on the plans. (This expansion joint could have been something as simple as some folded tar paper stuck between the slab and the foundation or stem wall.) Nothing serious has happened to the house, but a small crack did develop in the slab going from a corner of the hearth to the northwest corner of the house. When the tile was cemented down, it picked up the same crack. Marsh feels that this wouldn't have happened if he had used an expansion joint.

Laying the stone

Once the slab was poured and had cured, Marsh prepared to lay stone in earnest. He put up a set of boards at each corner of the building to which he could rig the string lines that would keep the outside of the 2-ft.-thick walls plumb. He calls them batter boards, but they're more like the story poles that some adobe builders use. They consist of a 10-ft.-long 2x4 post on the downhill corners and shorter 2x4 posts on the backfilled upper corners. The posts are plumbed in place and braced with diagonals, and lines are strung from one post to the other. Marsh's other tools consisted of a large 10-in. mason's trowel, a couple of smaller trowels including a 6-in. tool for pointing, and a 5-gal. bucket, shovel, hoe and wheelbarrow for mixing mortar. After some experimentation, he decided on a mortar mix of one-half bag of cement to 15 shovelfuls of sand and one shovelful of fireclay. He used the bucket for a water measure and learned to work a fairly stiff mix. (Marsh emphasized to me the importance of using a stiff mix for strength, but added that he was constantly tempted to use a wetter mix because it was so much easier to make and spread.)

With the reinforcing wire down over the sand and plastic membrane, and the stem wall laid, the plywood chute is ready for the floor slab to be poured. Photo by John Marsh.

The gravel for the drain system and the edge of the membrane as it comes up the back wall are visible here. Marsh used the string tied to the batter boards as a gauge for ensuring that the stones on the outside of the wall would be plumb. Photo by John Marsh.

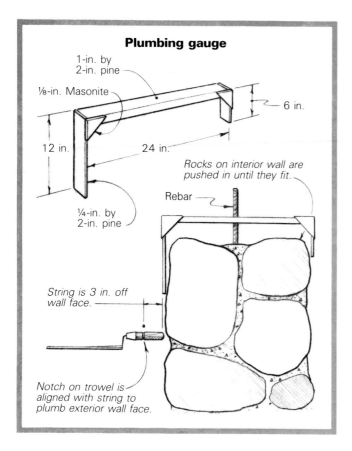

Plumbing gauge

1-in. by
2-in. pine

⅛-in. Masonite

6 in.

12 in.

24 in.

Rocks on interior wall are pushed in until they fit.

Rebar

¼-in. by
2-in. pine

String is 3 in. off wall face.

Notch on trowel is aligned with string to plumb exterior wall face.

both sides to make vertical joints. With his trowel handle, he measured 3 in. from the string to the face of the stone, then tapped the stone down with the handle of the trowel to get rid of any bubbles in the mortar. He made sure that the stones sloped to the outside of the wall to protect it from moisture penetration. Before moving on to the next stone, he made one more check against the trowel-handle notch. Next he test-fitted a stone for the inside face of the wall, put down the mortar and set the stone. Using the wooden gauge to check its position, he then tapped it into place.

Marsh next selected a stone to fill the space between the inside and outside stones and carefully mortared it in place. Some builders use rock chips and a minimum amount of mortar to fill the inside of a wall, but Marsh figured that a space filled with rubble would be a good place for moisture to accumulate and a wall built that way would be perpetually damp. A rubble-fill wall also can be weak structurally, because the inside and outside stones act as two separate walls. Some builders use tie stones to tie the inside and out-side of the wall together, but Marsh didn't find any 2-ft.-long stones, and if he had he probably wouldn't have been able to lift them alone. So he depended on his system of making a solid center to tie the wall together, using longer stones to span as much of the wall thickness as possible. Marsh viewed the total wall as if it were one of the dry walls his grandfather had built in New England, laying the stone as if mortar wouldn't really be holding everything together. To him, the placement of the stones in the middle of the wall was as important as anywhere else, although he did use the ugly rocks from his stockpile and the ones with freshly frac-tured faces there.

As Marsh worked, he bent the rebar around the rocks and worked the mortar up against the steel. As soon as he had used up a batch of mortar, he sprayed the work with a hose and covered everything with cardboard, sacks or plastic. It often took several days to complete one course around the entire house. Marsh measured his progress in wheelbarrowfuls of mortar used rather than in feet of wall laid, and figures he probably got 3 ft. to 4 ft. of horizontal wall out of one load of mortar.

As Marsh set each rock in place on the outside of the wall, he lined it up to the string by checking a notch he had made 3 in. from the end of the handle of the 10-in. trowel. For his inside plumb, he laid the wooden gauge shown in the drawing above across the wall and aligned the inside faces of the stones to it. The only real problem Marsh had was that he had to restring the lines each morning because ani-mals would chew through them every night. (As the walls grew past 4 ft. high, he added scaffolding and continued to work from the outside.)

Because Marsh had no stone-laying experience before starting Stonehaven, it took him a while to get the hang of it. His first two courses of stem wall do look a little rough. At the sides and back of the house most of the work is covered by backfill, but up front the pointing is uneven—some of it is up to the face of the rocks, some is recessed. In a few spots, the mortar had hardened before Marsh could even get back to point it. Higher up on the walls, however, there is a nice contrast of color and texture, and the rock sizes are balanced. The pointing is consistent and professional.

Before laying the stone, Marsh wet down the area of the wall where he would be working so that the water wouldn't be sucked out of the fresh mortar by the dry stone and mor-tar. Working from the outside of the wall, he test-fitted each stone, then spread on a mortar bed at least ¼ in. thick. He set the stone in the mortar, often using little pebbles—he calls them dobes—to prop up a stone or to get a better fit. The dobes were then thoroughly covered with mortar, so they wouldn't have to bear the weight of the rocks above all by themselves. Marsh then quickly troweled off any mortar that had oozed out, and tossed it alongside the stone on

Marsh laid up the walls amid a forest of vertical ⅜-in.-dia. rebar. This earthquake precaution weaves around the stones from footing to top plate on 18-in. centers, and ties to hori-zontal bar every 18 in. to 24 in. Photo by Kathryn Marsh.

Depending on how dry the day was, he would begin pointing his work an hour or two later; sometimes in cooler weather he could wait until evening. It took 20 to 30 minutes to point three wheelbarrows' worth of mortar. Many a time he found himself doing the work by the light of a kerosene lantern—he would wake up about ten o'clock at night realizing he hadn't yet pointed the mortar, so he would pull on his boots, light the lantern and go to work.

If the network of interlocking rebar that ties the walls, fireplace and pantry together were visible, it would look like a large cage. Most of this steel is tied into the center of the wall, with the stones worked around and between it. In addition to the ⅜-in.-dia. rebar on 18-in. centers used vertically, Marsh used more rebar of the same diameter horizontally on 18-in. to 24-in. centers. The poured concrete headers for doors and windows tie directly into this grid, as do the two main beams that support the loft. As noted earlier, the doors were braced in place before the slab pour; the window assemblies—jambs, sills and headers—were set in place and braced when the walls were high enough, then the stone was laid around the doors and windows. The sloping sills were poured in place with temporary formwork. The window headers were done a little differently. Marsh used fir 4x12s held flush to the outside and inside of the stone wall as a form for a concrete lintel. He lined this trough with plastic and ran rebar through from the walls on either side. After the concrete was poured, he left the 4x12s in place.

As the back walls rose above the top of the backfill, Marsh finished off the membrane system. He brushed two coats of fiber-laced asphalt emulsion on the stones, then unrolled the remainder of the plastic sheet and stuck it to the emulsion. After trimming the plastic to match the slope of the earth along the side walls, he created a shoulder of tar along the top edge of the plastic to ensure that no moisture would creep down the walls and behind the membrane. Before backfilling, he protected the membrane with some sheets of plywood as he shoveled in more drain gravel. As the gravel settled, he eased the plywood up the wall and out.

Scaffolding around the outside of the building allowed Marsh to lay the walls to 8 ft. high, at which point he stopped laying stone long enough to frame in the loft. The two floor areas of the loft are each 15 ft. by 10 ft., with a 4-ft.-wide connecting walkway along the north wall. The loft itself is supported by two main 4x12 beams running from the front to the back of the house. Marsh decked the loft with 2x6 tongue-and-groove pine, and then bridged in the open section and covered everything with plywood for a temporary work surface. This made it easier to ramp the stones straight across from the upper hillside (a section of the pantry wall had been left open for this purpose), and to lay stone from inside the house. The string lines outside the building were abandoned—the posts would have been too rickety at this height—and the loft provided a much better work area than scaffolding. Marsh then devised the adjustable-height sliding gauge shown at right to give him a measure of wall thickness and plumb at the same time. The gauge was designed to move along a 2x4 track on the loft floor; its rigid front edge kept the inside of the walls plumb, and the arm on top hung out 24 in. to keep the outside of the walls in line.

Here the sloping concrete sills are cast in place for the kitchen window on the left and a small window on the right, and the beams that will form the loft floor are temporarily held in place so the mortises for the joists can be marked. Photo by John Marsh.

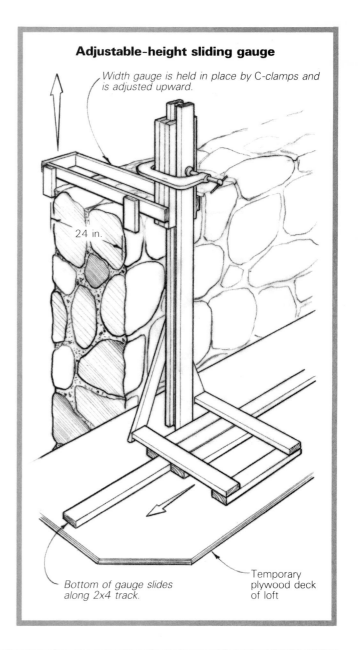

Adjustable-height sliding gauge

Width gauge is held in place by C-clamps and is adjusted upward.

24 in.

Bottom of gauge slides along 2x4 track.

Temporary plywood deck of loft

The roof

Before the rafters could be set, the walls had to be finished off. Six inches below the top of the walls, Marsh ran another course of horizontal rebar. He hooked J-bolts under the bar, and laid rock around them. A 2x8 redwood sill plate was bolted and mortared in place all around the perimeter of the building. Marsh figured his roof slope so that the windows in the east and west gable ends formed a diamond shape based on four 30°/60° triangles. This is roughly a 7-in-12 pitch.

Marsh found a mill in Lodi, California, that would custom-mill the 4x12 roof rafters. He set the rafters on 4-ft. centers and joined them at the peak with a 2x14 ridgeboard. On the south side of the house, the rafters run out beyond the stone wall, creating a covered front porch. Here they are supported by 4x4 posts. The roof decking is 2x6 tongue-and-groove pine. Over this Marsh put layers of 30-lb. roofing felt, 1-in. Celotex insulation board, more felt, insulation and a final layer of felt. A skylight stradles the roof at its midpoint.

To keep expenses down and to create a natural look, the architects had originally planned for a sod roof, but when Marsh heard that the 1920 administration building at Calaveras High School was being torn down, he inquired if he could salvage some of the roof tiles. He found that $450 would buy all he needed, and that was the last anyone heard about a sod roof.

The seven tons of beautifully colored tile turned out to be a kind of roll tile called S-tiles. They lie on the roof with one curve up, forming a dish, and the other curve down. One tile in a course overlaps the dish on the next. The tiles are nailed to the roof deck through predrilled holes.

The rakes and ridge called for some variations. Any gable roof has both left and right rakes, and because S-tiles are directional—they can be overlapped on only one side—the rake treatment for each end of the roof must be different. On the right rake, the open dish of the S-tile ends a course. This requires two semicircular tiles called barrel, or cover, tiles to make the roof weathertight at the gable end. The one on the outside that curls over and rests on the face of the barge rafter fits under the barrel tile that overlaps the dish of the last S-tile. The first barrel tile is nailed underneath the lapped area, and both barrel tiles are mortared in place. This same course of tile will begin on the other end of the roof (a left rake) with a single barrel tile nailed in place so that it covers the top edge of the barge rafter and is lapped by the S-tile. This joint is also mortared. The ridge is simpler; the barrel tiles are nailed to the 2x6 nailer, and all of the voids underneath are filled with mortar.

Marsh nailed down a 1x2 cant strip along the eaves to elevate the first course. He formed a double nailer with a 2x2 and a 2x4 on edge nailed through the insulation into the sheathing and barge rafter. Along the ridge, he nailed a 2x6 on edge. All these boards were covered with roofing felt. Marsh then cut all the roof flashing and built a little saddle (or cricket) for the upslope side of the chimney.

He next snapped a gridwork of lines on the felt. With the tiles piled in six stacks so as not to overload any part of the roof, he dry-laid a straight line of tile up the slope of the roof to get his spacing, so he wouldn't have to do any cutting. Finally he began nailing down the tiles, using galvanized nails and spacing the courses with 10 in. to the weather.

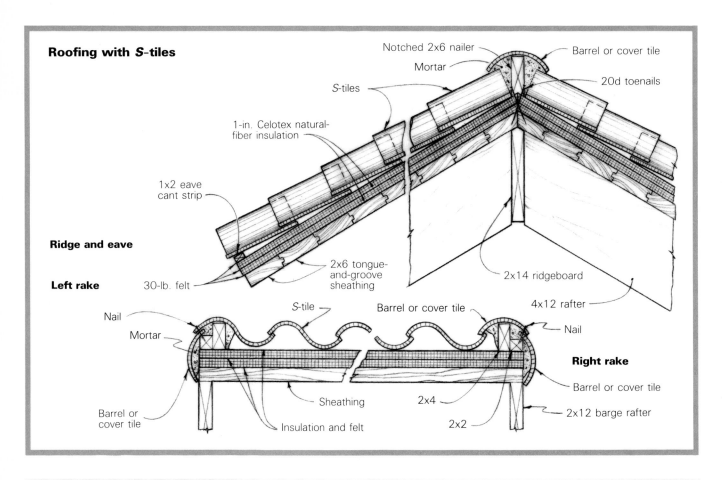

Roofing with S-tiles

Notched 2x6 nailer · Barrel or cover tile · Mortar · S-tiles · 20d toenails · 1-in. Celotex natural-fiber insulation · 1x2 eave cant strip · **Ridge and eave** · **Left rake** · 30-lb. felt · 2x6 tongue-and-groove sheathing · 2x14 ridgeboard · 4x12 rafter · Nail · Mortar · S-tile · Barrel or cover tile · Nail · **Right rake** · Barrel or cover tile · Sheathing · 2x4 · 2x2 · Barrel or cover tile · Insulation and felt · 2x12 barge rafter

The floor tiles

Kathryn Marsh was in charge of manufacturing the 3,000 hexagonal clay floor tiles for the house. The tools and procedures that John designed for the job are similar to those you would use for stamping out Christmas cookies, but the job itself proved to be a little more demanding and sometimes seemed endless—Kathryn worked through 1½ tons of red-stoneware high-temperature clay.

The clay purchased from a potter's supply store came in 25-lb. blocks, so Kathryn began by slicing it into 1½-in.-thick to 2-in.-thick slabs with a wire clay cutter. She then rolled each piece out on a muslin-covered work surface bordered by parallel 1x4s. By resting a kitchen rolling pin on these 1x4s and laying a slab of clay between them, Kathryn could roll the clay to a uniform thickness of ¾ in. The muslin kept the clay from sticking to the work table.

For cutting each tile to shape, John made a hexagonal cutter from a piece of 1x redwood. The hexagon is 6 in. across; the sides are equal and the angles are 90°. To form the cutting edge, he attached six pieces of aluminum roll stock to the redwood with brads and white glue. Each piece of aluminum overlaps the next by about ³⁄₁₆ in., and all are about 1 in. long. A 1x2 pine handle is attached to the top of the redwood with spacers, chamfered on all four edges for comfort, and drilled, countersunk and through-bolted to the redwood base.

After Kathryn punched out each tile, she collected the trim scraps and kneaded them back into blocks to be recut. The newly stamped tiles, still damp, were laid out on 4x8 sheets of plywood to dry in the sun. On hot days the tiles had to be turned every hour so they wouldn't cup. Even so, many of them had to be reflattened as they were turned.

The Marshes used a neighbor's kiln to fire the clay, and laid in 160 tiles at a time. The procedure that seemed to work best was a 25-hour firing that included 5 to 10 hours of preheating and 15 to 20 hours of high-temperature firing. In the kiln, the tiles shrank to their finished dimension of 5½ in. across and ⅝ in. thick.

The Marshes laid the tile in one weekend, using a creamy mortar mix on the slab. They began with a line of tile run right down the middle of the house from the fireplace to the middle of the pantry and leveled this first row carefully with a level fastened to the top of a 12-ft.-long 1x6. The rest of the floor was then leveled from this starter course.

To cut the edge tiles and filler pieces, the Marshes used a diamond-blade saw run off a generator. All the tiles were soaked in tubs of water just before they were laid, and the finished floor was allowed to set up for two weeks. John and Kathryn brushed in the brown grout with an old broom and cleaned up with sponges, vinegar and water. Finally, the floor was finished with a terra-cotta sealer.

Although the Marshes moved into Stonehaven once the tiles had been laid, they didn't complete some of the finishing touches for many months. They had also planned to build a small stone cottage next to the main house with a guest room, bath and laundry, but never found the time, so they had a conventional stuccoed building put there instead. It too has a tile roof and is tied to the stone house by the tile-covered front porch.

To make the hexagonal floor tiles, Kathryn Marsh cut each 25-lb. block of clay into manageable tile slabs, then flattened each piece with a rolling pin (top). Parallel 1x4s kept the tiles uniform in thickness, and muslin underneath prevented them from sticking to the plywood. Using an oversized hexagonal cutter, she then stamped out the tiles like cookies (above). Photos by John Marsh.

The main house has been very comfortable over the years. Although the fireplace worked well from the beginning, once Marsh built his separate studio up on the ridge, he realized that an airtight woodstove was more easily tended and heated the house more efficiently. Now the mouth of the fireplace is used for storing wood, and the little airtight sits on the hearth. One other change was ultimately welcomed—electricity finally came to Lupine Hill.

In the 14 years the Marshes have lived in the house, there have been two storms that Marsh remembers during which wind-driven rains buffeted the west wall for several days without letting up, and moisture made it all the way through the 2-ft.-thick wall. He says that if it happened any more often than that, he would hit the outside of this wall with a clear silicone sealer.

The House at Esperanza Creek

CHAPTER 3

Bob and Sandy Brabon's stone and frame house was a long time in coming. About 16 years ago, they moved from their home in San Diego to the Ebbets Pass area in California's Sierra Nevada mountains, where they spent their first winter in a 40-ft. trailer. The next summer, as they were looking for a better place to live, they heard about some summer cabins that were being torn down by their owners as the Forest Service cleared the private residences out of the Lake Alpine area. Brabon managed to save two octagon-shaped ones and a rectangular one from demolition and spent $1800 to move them down to Bear Valley. There he put an octagon on each end of the rectangle and covered the whole structure with shakes.

At the time, Brabon was a cook at a ski resort and knew more about French fries than about housebuilding. Sandy remembers that as they were putting the cabins together, Bob would invite the carpenters he met in the restaurant home for a beer and then ask pointed questions about construction. It wasn't long before he was doing more carpentry than cooking. This led to his first work in stone: a fireplace for the new house. Brabon then picked up another cabin transplant that had been damaged in the winter snow and with a makeshift crew put it back together. In addition to collecting cabins from Lake Alpine, he was collecting the skills he would use for the stone house, and just coincidentally he was building the equity he would need to purchase property and materials.

The snow gets deep at Bear Valley and it stays a long time. As they read and talked about what they wanted to do and where they wanted to be, the Brabon family (by now there were son Boone and daughter Dove) realized that their dreams of gardening and self-sufficiency would be harder to realize in the up country than at a lower elevation. The snow had been fun during the first couple of seasons, but the novelty of unplowed roads was beginning to wear off. In their reading, they came across Helen and Scott Nearing's book *Living the Good Life,* and developed a picture of what they wanted—a year-round stream, gardens, greenhouse, pond and stone house. The Nearings' story gave them the confidence they needed, and the method of poured-in-the-form rock building seemed sensible and quick. When they found the 40 acres on Esperanza Creek, its bed full of wonderful building stock, they sold the renovated cabins and made their move.

Brabon found steady work as a carpenter in Sonora, a two-hour drive from the new property. He began to build the homestead in the time left over after his 12 hours away each day. There was a primitive hunting cabin on the property; it had no windows, doors, bathroom, electricity or phone. It took three months to make the cabin livable, but it would be six years before the Brabons moved into their stone house in June 1979. Up above the cabin on a steep hillside was a level terrace where the previous owners had planned to build their house. There were even the beginnings of a poured foundation. It was there that the Brabons decided to build their home.

From the outside, the Brabons' California mountain house looks like a lot of two-story, gable-roofed, frame farmhouses with its country porches and second-floor wood siding. It's hard to see stone unless you look carefully. But when you do, you notice the 14-in.-thick stone walls on the first floor, made of river rock gathered from the creek that runs through the Brabons' property. On the south side of the house, the attractive stonework is clearly visible through the glass of the 40-ft.-long greenhouse. On the north side are a carport and a cold-storage cellar for the prodigious output from the family's garden. Porches extend from the east and west walls on both floors. The back porch (east wall) is enclosed to make an office, but the others are open.

Bob and Sandy Brabon built the first floor of their house using poured-stone construction, which on the inside of the house is finished with conventional 2x4 stud framing. Spanning the south wall is this 480-sq.-ft. greenhouse, where the Brabons grow most of their own vegetables. The cedar-encased door opening leads to the dining room.

The second floor is standard 2x6 frame construction with 1x10 cedar siding. Porches extend from both the east and west walls.

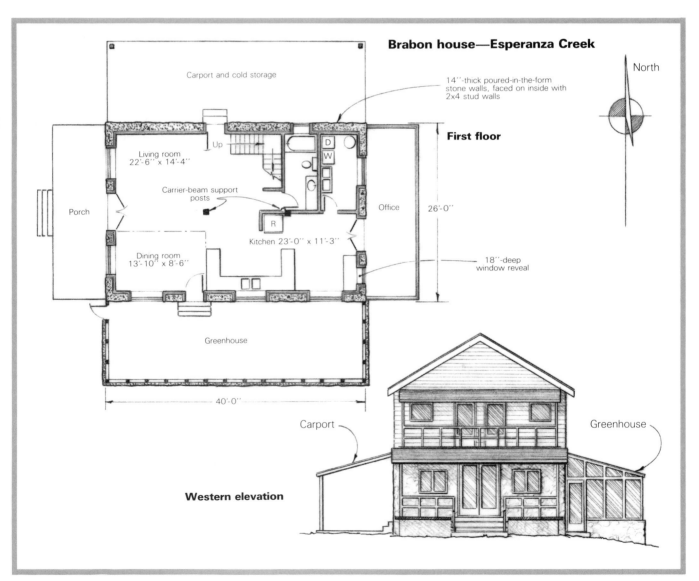

Brabon house—Esperanza Creek

North

Carport and cold storage

14''-thick poured-in-the-form stone walls, faced on inside with 2x4 stud walls

First floor

Up

Living room
22'-6'' x 14'-4''

D

W

Carrier-beam support posts

Porch

Office

26'-0''

R

Kitchen 23'-0'' x 11'-3''

Dining room
13'-10'' x 8'-6''

18''-deep window reveal

Greenhouse

40'-0''

Carport

Greenhouse

Western elevation

The stonework on the first story of the Brabons' house was done with a poured-in-the-form system similar to the one described in the Nearings' book. Most builders who use the poured-stone method leave the stone exposed on the outside walls (the face stone) and pour straight concrete up against the forms for the inside walls. (It's possible to have face stone on both sides of the wall, but this entails a lot of extra work—you have to place the stones in the forms carefully, and have a greater number of stones with attractive faces on hand.) The beauty of the poured-stone system is that the work goes quickly, and broken and ugly rock can be used as filler behind the good-looking face stones. But with the low insulative value of stone and the possibility of moisture penetration, many builders, the Brabons included, feel the need for a vapor barrier and insulation on the inside of the walls. The flat concrete interior surface of a poured-stone wall makes this an easy matter. With a vapor barrier and a sheetrock-covered 2x4 stud wall inside the stone wall, the interior of the Brabons' home looks very much like a conventional frame house. It is not until you notice the shelf-like windowsills that you realize there is something beefy beneath that plain exterior.

The north wall, under the carport roof, shows a nice variety of texture and size of stone. The wall was formed and then later pointed with mortar after the house was finished.

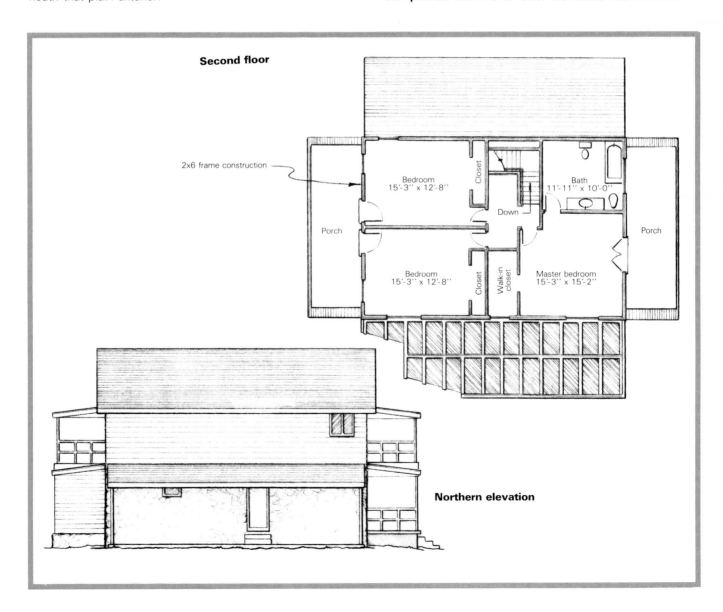

Second floor

2x6 frame construction

Porch

Bedroom
15'-3'' x 12'-8''

Closet

Down

Bath
11'-11'' x 10'-0''

Porch

Bedroom
15'-3'' x 12'-8''

Closet

Walk-in closet

Master bedroom
15'-3'' x 15'-2''

Northern elevation

Getting started

As the Brabons were designing their house and drawing up plans, they brought in a Caterpillar tractor to enlarge the terrace so there would be enough level area for parking, a workshop, outbuildings and a garden. They also had the tractor dam up the ravine west of the house for a pond.

The house was built with the slip-form system discussed in Chapter 1 (pp. 12-14). In this system, a pair of forms is wired together on opposite sides of the foundation wall and then filled with stone and concrete. When the forms are filled and the concrete has cured, another set is put on top and filled. Then the bottom form is stripped and put on top, and so on until full wall height is reached.

Brabon decided on a basic form size of 2 ft. high by 8 ft. long, although some were shorter. He built the forms out of 2x4 frames and ¾-in.-thick BB plyform plywood. The forms are held together with tie wire. He cut all the 2x4 bracing about ⅛ in. short so that the plywood would stick out slightly. This way, the forms would butt tightly together even if the rails were a little bowed. While this was probably unnecessary for the stonework of this house, Brabon has found that it's helpful in making good-looking poured-concrete foundations. (He now builds custom homes in Calaveras County, California. He saved all the forms he used on his house and now uses them in his foundation pours. In fact, he has done several of the foundations for David Easton's earth houses, including the Sturgeons' house discussed in Chapter 10.)

Brabon made enough forms to go around the house twice. He spent about $1,000 on wood for the forms, and he says that building them was fairly easy. Because they were all basically the same, he could cut all the wood at the same time. When he finished the forms, he poured the foundation.

The Brabons designed the house so that the north and south walls could take advantage of the existing foundation on the property. The footings and stem wall weren't big enough to support the 14-in.-thick stone walls, but Brabon was able to increase the foundation's strength by digging down and expanding the footings. (He then used the stem as a ledge to support the floor joists.) On the east and west walls, Brabon poured a 3-ft.-wide by 16-in.-thick footing with no stem wall. On these walls, he would start the stonework right on the footings, but on the north and south walls, he would build up the stone on the outside of the stem wall (drawing, facing page). Three pieces of ½-in.-dia. rebar were run horizontally through the footings, and ½-in.-dia. vertical rebar was set on 24-in. centers to extend up through the stone walls. Brabon inset the vertical rebar 10 in. from the outside of the walls so that it would be out of the way of the large face stones. He also ran five courses of ½-in.-dia. horizontal rebar all around the house on 24-in. centers, which when tied to the vertical bar created a grid of steel throughout the walls. This was the same idea as the system John Marsh used (Chapter 2) and is required by code in most parts of earthquake-prone California.

As the Brabons gathered their building material from the creek bed, they looked for stones in three categories: face stones, corner stones and filler stones. The face stones for the outside of the wall had to have one nice-looking surface and a flat bottom that would nest into the wall securely. The corner stones needed not only a flat bottom but also two good-looking faces square to each other to define the corners. The filler stones would be used to fill in between the other stones and to take up space on the inside of the wall where the bulk of the concrete would be poured.

The Brabons filled their pickup truck with stone until the springs were groaning and then drove it up from the creek to the house site. They learned early not to heave the stone out because that was a good way to damage the face stones. Brabon figures they lifted each stone at least seven times: out of the creek and into the truck, out of the truck and onto the ground, into the wheelbarrow, out of the wheelbarrow and onto the scaffolding or top edge of the form, into the wall for a test fit, out of the wall so the concrete could go in, and back into the wall for good.

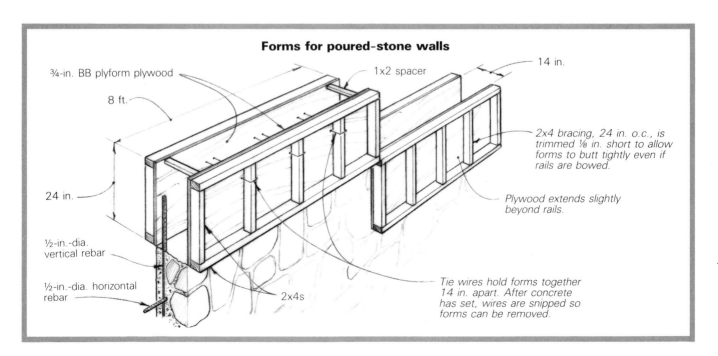

Forms for poured-stone walls

¾-in. BB plyform plywood

8 ft.

1x2 spacer

14 in.

2x4 bracing, 24 in. o.c., is trimmed ⅛ in. short to allow forms to butt tightly even if rails are bowed.

Plywood extends slightly beyond rails.

24 in.

½-in.-dia. vertical rebar

½-in.-dia. horizontal rebar

2x4s

Tie wires hold forms together 14 in. apart. After concrete has set, wires are snipped so forms can be removed.

Section of north/south wall

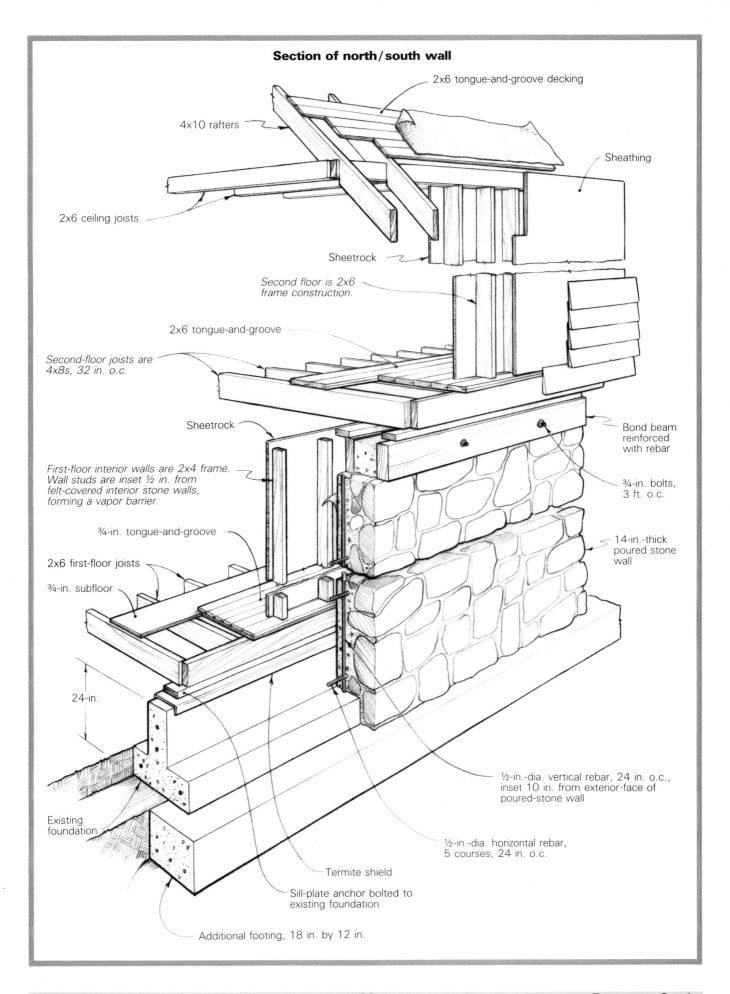

2x6 tongue-and-groove decking

4x10 rafters

Sheathing

2x6 ceiling joists

Sheetrock

Second floor is 2x6 frame construction.

2x6 tongue-and-groove

Second-floor joists are 4x8s, 32 in. o.c.

Sheetrock

First-floor interior walls are 2x4 frame. Wall studs are inset ½ in. from felt-covered interior stone walls, forming a vapor barrier.

¾-in. tongue-and-groove

2x6 first-floor joists

¾-in. subfloor

Bond beam reinforced with rebar

¾-in. bolts, 3 ft. o.c.

14-in.-thick poured stone wall

24-in.

Existing foundation

½-in.-dia. vertical rebar, 24 in. o.c., inset 10 in. from exterior-face of poured-stone wall

½-in.-dia. horizontal rebar, 5 courses, 24 in. o.c.

Termite shield

Sill-plate anchor bolted to existing foundation

Additional footing, 18 in. by 12 in.

Building the walls

After the foundation had cured, Brabon installed the first course of forms around the entire perimeter of the house. On the east and west walls, where forms were placed right on the footings, he used 1x2 wood spacers to hold the forms apart, removing them as he filled the forms with stones. On the north and south walls, he used the stem wall as the inside form and braced the outside form so that stone could go all the way down to the footings, just as on the east and west walls. (The 24-in.-high forms were planned to match the existing foundation.) Where two forms butted together end to end, he nailed the adjacent 2x4s with duplex nails.

The system you use to build the walls is determined by the number of forms you have. With just a few sets, the forms must be moved laterally around the building as you work, as well as leapfrogged vertically. Ideally you should do as much wall as possible at one time to minimize the cases where a newly poured joint meets an already cured one. Although the Brabons formed up the entire building to 24 in. high, they didn't have the time or manpower at first to work more than 10 running ft. at a session. Sometimes they worked only 5 ft. or 6 ft., but Brabon wasn't worried about cold joints in the concrete because he left plenty of stone sticking out on the ends to tie in with the next pour.

Because the outside dimension of the walls is longer than the inside dimension, a couple of sets of short forms were needed. The outside dimension of the north and south walls (40 ft.) was covered by five 8-ft.-long sections, but the inside (37 ft. 8 in. long) needed four 8-ft.-long sections plus a filler measuring 5 ft. 8 in. in length. For someone working with only several pairs of forms (that is, without a full set), this difference in dimension would not be a problem, as the forms could be staggered at the corners. For Brabon, forming the entire perimeter at one time as he did, the short forms were necessary.

In poured-stone walls, concrete glues the stones together. Mortar is used, but only cosmetically—after the forms are stripped, mortar is used to fill in the exposed openings between stones. Brabon made his concrete from one part cement, four parts ¾-in. gravel aggregate and three parts sand, adding just enough water to make the mix work easily between the stones, but not so much that it would run out over the surfaces of the stones. The mix was stiff enough to stand up—about a 3-in. slump. Brabon mixed the concrete in a gas-powered mixer and used a wheelbarrow and 5-gal. buckets to deliver it to the walls. He used 100 bags of cement, lots of ¾-in. gravel and a good-size pile of sand before they were done.

Stilts made from 2x4s help hold a double row of 24-in.-high forms in place. The holes in the wall are under the window bucks (rough jambs) and will be filled with small rocks and mortar later.

To start, the Brabons poured about 2 in. of concrete from the buckets into the bottom of the forms. Bob set the face stones into the wet concrete, their faces tight against the outside form board. During the work, he wore heavy rubber gloves, working the concrete in around the stones with his fingers. When the stones were firmly seated, he test-fitted the next course, Sandy shoveled in more concrete and he set the stones in place. He tried to choose stones that were fairly flat on the top and bottom so they would bed securely: (Even with formed stone, it's a good idea to lay the stones so that they will stay in place even without mortar.) Then he tucked in the smaller filler stones around them, making sure there was at least 1 in. of concrete between all the stones. Some filler stones were used to prop up the backs of the face stones so they would stay tight to the form board. Others were simply used to take up space in the 14-in.-thick walls so they wouldn't be solid concrete.

As the Brabons worked the stone, they also had to work around the vertical rebar. But because it was inset 10 in. from the outside face of the wall, this was not as hard a job as it was in the Marsh and Monson houses (Chapters 2 and 4, respectively), where the rod constantly had to be bent around the stones. The Brabons generally could work around the bar with filler stones. And with the rebar grid running on 24-in. centers, it was an easy matter to position the larger stones into one of the 24-in. squares created by the intersecting bars.

Every now and then Brabon used a tie stone, one that crossed the full 14 in. of wall thickness, but most of the inside of the wall was filler stone and concrete. As he reached the top of each course, he was careful not to top off the form with concrete, but instead to leave some stone exposed to tie into the next course. This is necessary in poured-stone construction to avoid having horizontal bands going around the house at each form juncture; instead, the walls should look as though they were laid up by hand, one rock atop another. The last thing Brabon did at the top of each 24-in.-high course was to tie in the horizontal rebar before adding the next set of forms.

The hard part of laying stone in a form is that you must develop a feel for what the wall face looks like, because you can't see what you've done until you pull off the forms. You even have to remember what rocks you've put in if you want an interesting variety of size, shapes and colors. And sometimes there are unpleasant surprises. At the beginning of this project, for example, Brabon poked the concrete in between the stones so zealously that it oozed out onto the outside faces of the stones. When the first forms were stripped off the bottom, the problem with his technique became obvious, but by that time he had already poured the first 4 ft. of wall. It fell to son Boone to break out the concrete using an electric jackhammer. Lesson learned, Brabon was thereafter careful not to push the concrete out to the exterior face and found that he could clean up the occasional spot that oozed out by knocking it away with a hammer. (It was several years after they had moved into the house before he got around to finishing up the walls by mortaring the joints between the rocks. When Brabon did finally get to

mortaring the voids between stones on the outside of the walls, he mixed up a rich mortar of 2½ shovelfuls of sand to 1 shovelful of cement, adding enough water to make the mix nice and creamy. First he chipped out all the loose concrete, then he put the mortar in a mortar bag and squeezed it into the cracks the same as you would frost a cake with a pastry bag. He then pointed the joints with a small trowel and washed the rock with acid before wire-brushing it all clean.)

When the first course of wall had cured, Brabon nailed a second set of forms to the first set with duplex nails through the bottom rail of the top form into the top rail of the bottom form. The Brabons worked the stone alone until they got to the bottom of the windows. This was as high as Sandy could lift the stone, so Bob decided to hire some taller helpers—Sandy's brother, Duncan, and their neighbor, Hoadie.

Brabon isn't afraid of big stones. He proudly points out some of the 300-pounders that taxed everyone's imagination. As the walls grew and he kept picking out monster stones, the crew rigged two scaffold jacks that could be pumped up by foot to raise and lower the scaffolding like an elevator. Now they could put the wheelbarrow full of concrete on it, add some giant stones and pump it up close to the top of the wall.

Boone Brabon breaks oozed-out concrete off the face of the stone with an electric jackhammer. His father's technique improved after the initial pour.

Looking into the forms, you can see the heavy 1⅛-in.-dia. rebar used over the door and window openings. The forms are wired together, and the wooden spacers are removed as the rocks and concrete are placed in.

The concrete bond beam solidly links all the stone walls together at the top.

Where Brabon wanted a water faucet on the outside of the stone wall, he put some ABS pipe in the forms to make a hole through which the water pipe could be slipped. When he reached the proper height for the vent openings and the window and door frames, he built them of 2x14 cedar and set them in the forms. He wrapped the cedar in 30-lb. tar paper to protect the wood from direct contact with the concrete, and then drove nails into the outside of the wood to lock into the walls. The Brabons had picked up a number of antique doors over the years, so they had to size the openings to fit each particular one. Over the tops of the windows and doors, Brabon ran some #9 (1⅛-in.-dia.) rebar in the concrete headers. He tied this in to the ½-in.-dia. rebar network in the walls.

At the tops of the stone walls is the bond beam, formed with large pieces of lumber: a 4x10 on the outside bolts to a 2x10 on the inside with ¾-in. bolts every 3 ft., doubled up on the ends and where boards splice together. Brabon would use the 4x10 to anchor the rafters for the porches, carport and greenhouse he planned to add later. He then pumped the concrete for the bond beam with a concrete-pumper truck. (For a full discussion of bond beams, important to all types of building in this book, see Chapter 9.)

Finishing up

With the stone walls up, Brabon started work on the floor and the interior framing. He put down 2x6 floor joists on 16-in. centers—the centers of the joists were supported by a beam on piers up the middle of the house. After insulating between the joists, he nailed down a ¾-in. plywood subfloor, which later would be covered with ¾-in. tongue-and-groove, vertical-grained, kiln-dried, clear fir. Underneath the floor there is a crawl space with access through a crawl hole in the stone wall.

Brabon then built what amounts to a frame house inside the stone walls, to cover the bare concrete of the inside of the formed-stone walls. First he laid on 30-lb. builder's felt for a vapor barrier on the inside of the stone walls, and then he built the 2x4 stud walls, setting them out ½ in. from the felt for some dead airspace. Wiring and plumbing went into these walls, and then Brabon packed them with insulation and nailed on the sheetrock. By doing the walls this way, he has avoided the problem of dampness usually associated with stone houses, but he also forfeited much of the thermal advantage of the mass of the stone. In the winter the heat inside the house does not warm the stone, nor does the stone cool the interior in the summer as it might if the masonry were exposed. (Studies of high-mass building materials suggest that it's better to apply the insulation to the outside of the walls for the full thermal advantage of the materials, but the Brabons wanted to see their stone on the outside.) Brabon has been happy with the way the house heats and cools, however. He uses very little wood in the winter for heating and no supplemental cooling during the hot summer season.

A huge 4x12 beam divides the Brabon house in half lengthwise as it supports the floor system for the second story and the roof system above. Two 4x8 posts support this beam. During one California earthquake, Brabon observed the 4x8 post in the living room flexing and twisting enough to make him a little nervous. To ease his own fears, he bolted a couple of 4x6s to the 4x8s with ⅝-in. bolts. Since then, he hasn't had the opportunity to observe them during an earthquake, but they look strong and now have a nice, sculpted, craftsman look.

The exposed 4x8 second-story floor joists are on 32-in. centers and are covered by 2x6 tongue-and-groove decking that is glued down and blind-nailed. Brabon sanded it all down and then coated it with a commercial gymnasium-floor finish. The upstairs is all 2x6 framing with fiberglass insulation and 1x10 cedar siding on the outside.

Up top, the Brabons have 4x10 rafters covered by exposed 2x6 tongue-and-groove roof decking, 1½-in. Techni-foam rigid insulation and a weather roof of glazed concrete tile. On top of the rigid insulation, Brabon laid his roofing felt and then ran 1x6s parallel to the eaves. The tile was then nailed to the 1x6s.

Sandy Brabon designed the oak cabinets in the kitchen. For the chopping-block counter that divides the kitchen and dining areas, Brabon started with an 8-ft.-long piece of gluelam Douglas fir 6 in. thick and 24 in. wide. He covered the sides of the fir with oak and covered the rest of the countertops with ¾-in. oak flooring. On the doors and cabinet ends he glued some light-colored, Italian-hardwood decorative motifs. A centerpiece in the kitchen is the beautiful 1929 Wedgewood stove that came from another of the doomed Lake Alpine cabins. It's set up for either wood cooking or bottled gas.

On the north side of the house, Brabon added an above-ground root cellar. The carport roof covers it and one wall is the outside stone of the house. He framed it with 2x6s and insulated it with 6-in. fiberglass batts all around. Now the family keeps tomatoes, fruit and root crops in it through most of the winter.

The 12-ft. by 40-ft. greenhouse was the last big project to be completed on the house and adds even more to the Brabons' food production. Brabon picked up some 34-in. by 76-in. tempered sliding-door blanks for $15 each and installed them in 13 banks across the front of the house. Each bank has one vertical sheet on the front and two pieces end-to-end for the glass roof. The framework is made up of painted 2x4s, and inside there is a 40-ft. growing bed next to the windows and a 30-ft. bed up against the stone wall. This 480 sq. ft. of sun-heated area adds a lot of warmth to the house in the winter. In the summer, the doors on both ends keep it from getting too hot and overheating the house. Boone's cactus collection loves it and the vegetables grow all year round. The fig tree in front of the kitchen window thinks it lives in the tropics.

Here the stonework is complete except for the mortaring of the joints. The second floor is framed in with 2x6 walls, 4x10 rafters and exposed roof decking. Sandy Brabon is on the ladder.

The oak kitchen cabinets in the Brabons' house were designed by Sandy Brabon and built by her husband, Bob. One of the two main posts that support the carrier beam divides the living and dining areas.

Toad Hall

CHAPTER 4

Every inch of Toad Hall is an expression of the hands and hearts of Bob and Cara Monson, of Amador County, California. Built on the banks of a mountain stream, the house is situated below gardens, a greenhouse, and a pond with a waterwheel and pump (to feed the water back to the gardens and greenhouse). Its arched windows and doorways filled with beautiful wood doors and stained-glass windows made by Monson give the house a magical air.

The Monsons hadn't really planned to build a stone house. Bob Monson, a chemist, woodworker and violin maker, wanted a house that was sculpted and not just nailed together. Before building Toad Hall, the Monsons first put up several buildings on their property, including a woodshop where Bob would build the cabinets, windows and doors for the stone house, a blacksmith shop, a generator house and a cook house (now the pottery studio). They lived upstairs in the woodshop, cooked in the cook house and bathed at the greenhouse with a solar-heated shower. The complex took two years to complete; the house itself took another eight.

As Monson was still planning the main house, he added a basement to the little cook house, made from vertical timbers with stone fill between them. The stone looked so good that he decided to use it for the main house, too. In addition, stone seemed to suit the site, which used to house an old sawmill. The mill operated from 1926 until World War II, and in that time it created a hillside of sawdust that is still 30 ft. deep in some spots. (Monson moved the sawdust off the house site with a D-2 Caterpillar tractor that he found and rebuilt—the 40-year-old sawdust proved to be a hot

commodity with nurseries in the San Franciso Bay Area, and the Monsons traded it for many of the plants that now enhance the site.)

Cost was also a factor in the decision to build with stone, as Monson figured that stone would be free for the hauling. As it turned out, he did a lot of hauling. With all the sawdust, there wasn't much stone on the property, so Monson searched out free sources. (Most people who build with stone buy a piece of property with stone on it or at least nearby.) He found some nice limestone at several nearby quarries that crushed rock for road paving. Some of the rocks were 200-lb. and 300-lb. rejects from the rock crusher, and he had to use ropes, ramps and winches to handle these giants. But he got most of his stone from the south fork of the Mokelumne River in northern Calaveras County—about 18 miles from the building site—bringing it home in his old truck a ton and a half at a time. When he looks at his walls now, Monson jokes that the local rockhounds would tear the place down if they knew about the wonderful hunks of jasper and agate they would find. None of the stone is cut, nor did Monson shape it or break it as he worked. He wanted rough, weathered surfaces. When asked what he would do differently if he were to build another stone house, Monson says that for one thing he would go out with several men, point at the stones he wanted and let them do the lifting.

The interior of the Monsons' house is spacious—over 1900 sq. ft. (photos, pp. 40 and 46). The downstairs has a large kitchen and dining area at the north end, and the entry, woodstove and stairs in the center. On the south end are the yet-to-be-completed spa/bathing room with stone shower, and a bathroom with stone floor and pebble countertop. Upstairs there is enough room for three bedrooms, but the Monsons like things open and so there is one giant bedroom/sitting room/library/living room/music room that measures over 1,000 sq. ft. The trusses and purlin system (described on p. 46) are all exposed, and the hexagonal skylights are tiled around the edges with yellow and blue glass tiles.

Toad Hall, built of hand-laid stone set in concrete, is the work of Bob and Cara Monson. Most of the stone for this California mountain home was hauled from a riverbed 18 miles away. The second floor, cantilevered out over the stone walls, consists of one huge room that measures over 1,000 sq. ft. and is the couple's main living area.

The Monsons' dining area is centered around two Gothic-arched windows.

The main entryway is dominated by the center staircase leading up to the second-floor sleeping/living area.

Downstairs stone is everywhere. There are pebble kitchen countertops, pebble floors with drains in them (so they can be hosed down), and a stud wall with a pebble facade behind the woodstove. Cara Monson deserves the credit for their interest in pebblework. She has always been a pebble collector; on an outing to the beach she would spend the day gathering pretty stones and bring them home by the bagful. When they lived in Berkeley, the Monsons started to build pebble-top tables just to have something to do with Cara's collection. Soon they were doing the entry floor of their apartment building in pebbles, then the bathrooms, and then even putting pebble panels in the garden walls.

When they built Toad Hall, the Monsons used mortarless, hand-laid stone construction. The walls have open joints and a rugged, ancient quality. But though the word "mortarless" implies that there is nothing to hold the stones together (as in dry-laid stone walls), such is not the case. The stones in a mortarless wall are glued together with a concrete mix and gravel aggregate, rather than with the more usual cement mortar, which has no aggregate in it. There are several advantages to building mortarless walls. For one, fewer bags of cement per job are needed than for cement-rich mortar, so this method is less expensive. Another advantage is that the gravel in the concrete holds the stones in the wall apart and prevents them from squeezing the concrete out—a heavy stone can compress a cement mortar mix until there is not much mortar left between it and the stone below. (John Marsh, in Chapter 2, used pebbles to prevent this.) Because the gravel in the concrete holds the stones apart, the builder who builds mortarless walls can build higher vertically in one session; with cement mortar the builder must build horizontally and wait until the mortar has cured before adding courses. Builders who use poured stone (as the Brabons did in Chapter 3) use the same concrete as do mortarless builders to hold their stones together.

When you use the mortarless method, you must come back after the stones have been laid and point the joints with cement mortar. The concrete doesn't squeeze well enough to fill them. (It can fill the joints in poured-stone walls, but the concrete doesn't look like mortar, just like drippy concrete.) Pointing with a traditional cement mortar mix seals the joints and prevents moisture infiltration—in an area with heavy freezes, any water in the open joints could freeze and either break the stones or cause the walls to crack.

After saying that it's necessary to point the finished mortarless stone wall, I'll tell you that the Monsons didn't point theirs. They left the joints open, to the delight of the bugs and small creatures that have made them their homes. It doesn't bother the Monsons. They like the way the unpointed stone looks. They like the fact that they didn't have to do all the extra work involved in pointing, and that they saved the money it would have cost to buy enough cement mortar to do their large house inside and out. The Monsons have had no problem with moisture infiltration, freezing or cracking. Part of the reason is that the house has a generous roof overhang and covered porches on three sides. And while their 4,000-ft. elevation gets a fair amount of snow, they don't get the severe frost that they would at higher elevations or in colder parts of the country.

Although it's customary in mortarless construction to point the joints between the stones, Monson left these spaces unfilled for a rougher effect.

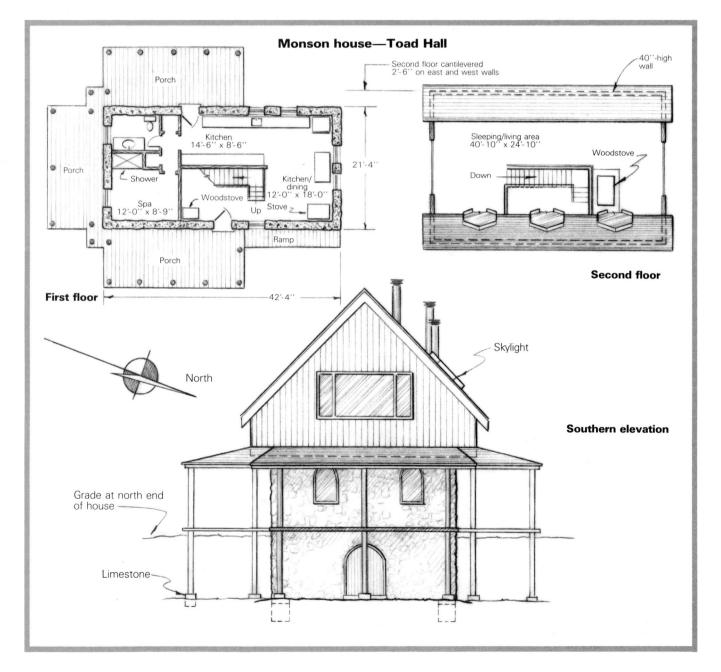

Monson house—Toad Hall

First floor

Porch

Porch

Porch

Kitchen 14'-6" x 8'-6"

Shower

Kitchen/ dining 12'-0" x 18'-0"

Spa 12'-0" x 8'-9"

Woodstove

Up

Stove

Ramp

21'-4"

42'-4"

Second floor cantilevered 2'-6" on east and west walls

40"-high wall

Second floor

Sleeping/living area 40'-10" x 24'-10"

Woodstove

Down

North

Skylight

Southern elevation

Grade at north end of house

Limestone

The foundation

Toad Hall is built on a hillside and the first floor is 6 ft. to 7 ft. off the ground on the downhill side. This creates a good storage space under the house and a convenient place to run plumbing and electricity. At the midpoint there is still a good 4 ft. of clearance under the floor. With the quantities of sawdust and spongy soil on the site, Monson wanted to fasten his heavy stone house firmly, so he hand-dug the footing trenches right down to bedrock (in most cases he had to go down only a foot or so, but in some spots he had to dig down 3 ft.). Then he washed the rock clean for a good bond to the concrete. Because Monson didn't want any concrete to show, he stepped the footings down the hill and built the stem walls that support the floor joists out of stone. For ventilation, the stone stem walls have pieces of 12-in.-dia. steel culvert set into them on 6-ft. centers.

The footings are 25 in. wide and were formed with vertical 1x12s set in the trenches and a 2x4 whaler nailed horizontally across the outside top edge and bolted at the corners. Monson ran two pieces of ½-in.-dia. (#4) rebar along the bottom of the footings and another two pieces 6 in. below the top. He used a system of tied rebar in the stone walls similar to that used by John Marsh in Chapter 2. As he prepared to pour the footings, he had to tie in the vertical runs of ½-in.-dia. bar for the stone walls. These 10-ft.-long pieces of bar were run in pairs about 6 in. apart every 3 ft. He also ran a pair up each corner of the building. Monson poured the concrete plinths for the wood posts that support the mid-span girders at the same time he poured the footings.

The stonework

The Monsons started gathering stone while they were still building their first shelter on the property. Like other stone builders, they looked for stones that were relatively flat on the top, the bottom and the face that would be exposed to the weather. Monson had thought that the hardest stones to find would be the ones for the arches, but these were actually fairly easy to come by. The ones that were most difficult were the corner stones, which had to be flat on both top and bottom and have two faces at 90° to each other. Because the stonework is also exposed on the inside walls of Toad Hall, the Monsons had to gather a lot more good-looking stone than builders like the Brabons (Chapter 3), who built a poured wall with concrete facing on the inside.

Continuing their unorthodox approach to stonework, the Monsons didn't even use mason's tools when laying the stone. They mixed the concrete in a cement mixer and brought it to the wall in a wheelbarrow, but they didn't use any trowels. Bob hoisted each stone into place for a test fit and then removed it. Cara then dipped into the concrete with a 1-gal. steel plant can and poured the concrete onto the top of the wall, keeping it away from the outside or inside edges. Bob then set the stone, placing the face even with the edge of the wall. He stepped back and sighted along the entire wall to make sure the new stone was in line, then wiggled the stone into place, flattening the concrete as he did so. The ¾-in. aggregate in the concrete kept the spacing between stones consistent. Monson was careful to run the lengths of the stones across the wall for more

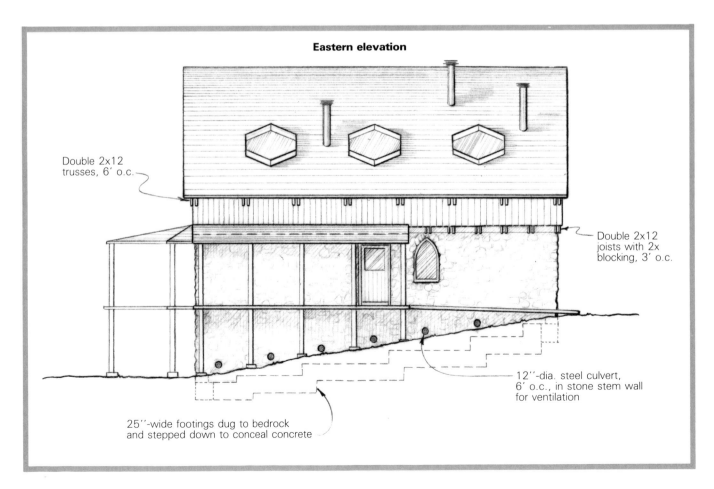

Eastern elevation

Double 2x12 trusses, 6' o.c.

Double 2x12 joists with 2x blocking, 3' o.c.

12''-dia. steel culvert, 6' o.c., in stone stem wall for ventilation

25''-wide footings dug to bedrock and stepped down to conceal concrete

strength, though the stones that actually bridge the wall from one side to the other are rather rare—he didn't come up with many 20-in.-long stones with nice faces on both ends. But by running the stone lengthwise into the wall from both sides, he created a series of interlocking fingers that give the wall maximum strength. After Bob had several rocks in place along the face of the wall he was working on, Cara dumped more concrete over their tops, and with a small stick Bob worked the concrete into the vertical cracks.

For the Monsons, working alone, a good day's work used two mixers full of concrete. At one point, a weekend neighbor and son of a professional mason came by to help, and that morning they were able to use four mixers full. After that, Monson decided that for his next stone house he'd hire a helper to keep the concrete coming to the walls.

The walls of Toad Hall are a full 25 in. thick up to floor level. At that point, the walls become 19 in. thick, creating a 6-in.-wide shelf along the inside of the walls on which to set a redwood sill. From the shelf on up, the walls vary in width from 18 in. to 22 in., the inevitable result of eyeballing alignments and working with uneven widths and lengths of stone— the Monsons didn't use batter boards or string to keep the walls plumb. A pair of 6x6 girders was set down the middle of the house with the 2x6 floor joists going across them and resting on the redwood sill. Once the 2x6 tongue-and-groove subfloor was nailed down, the Monsons had a platform to work on. The walls on the downhill (south) side of the house were 6 ft. high by this time, and the scaffolding was climbing, so the floor became a nice place to work. The level surface would support a ladder, boxes and simple scaffolds as the walls continued to rise. Although Monson built lots of scaffolding, it proved too narrow to fit stones, workers and a wheelbarrow of concrete—another aspect of the work that Monson would change were he to build another stone house. He would also use a scaffold jack to lift rocks and wheelbarrows to the top, as on p. 35, instead of his system of winches, hoists, and block and tackle. He figures that all the wood for such a system would probably cost a couple of thousand dollars, but that he could recycle it into the house as part of the roof system once the stonework was done.

Being subject to the same California earthquake codes as the other stone-builders in this book, Monson was required to run both vertical and horizontal rebar in the stone walls. In addition to the vertical ½-in.-dia. rebar at each corner and the other two pieces run every 3 ft. around the house, he tied a piece of horizontal bar to the verticals every 3 ft. Monson's building inspector had explained early on the need for a tie stone running all the way through in every 9 sq. ft. of wall. Since he couldn't come up with that many long stones, Monson had proposed using lengths of ⅜-in.-dia. rebar with a hook bent in each end instead. (The hook grabs the stone and concrete, preventing the rebar from sliding through.) The inspector approved, and Monson laid these homemade ties wherever they were convenient—he ended up using more than 200 of them. Like many owner-builders, Monson used more steel than was necessary, but he didn't want to have to worry about the strength of his walls. As he was finishing up the stonework, he ran a last horizontal piece of rebar through the wall all around the top of the house.

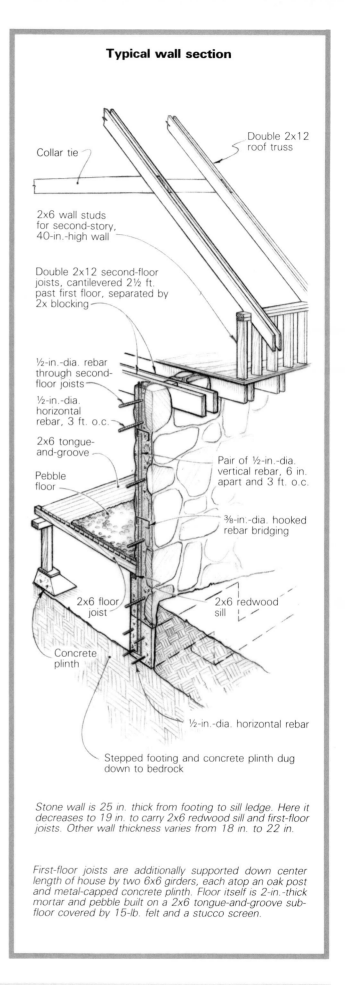

Typical wall section

Collar tie

Double 2x12 roof truss

2x6 wall studs for second-story, 40-in.-high wall

Double 2x12 second-floor joists, cantilevered 2½ ft. past first floor, separated by 2x blocking

½-in.-dia. rebar through second-floor joists

½-in.-dia. horizontal rebar, 3 ft. o.c.

2x6 tongue-and-groove

Pebble floor

Pair of ½-in.-dia. vertical rebar, 6 in. apart and 3 ft. o.c.

⅜-in.-dia. hooked rebar bridging

2x6 floor joist

2x6 redwood sill

Concrete plinth

½-in.-dia. horizontal rebar

Stepped footing and concrete plinth dug down to bedrock

Stone wall is 25 in. thick from footing to sill ledge. Here it decreases to 19 in. to carry 2x6 redwood sill and first-floor joists. Other wall thickness varies from 18 in. to 22 in.

First-floor joists are additionally supported down center length of house by two 6x6 girders, each atop an oak post and metal-capped concrete plinth. Floor itself is 2-in.-thick mortar and pebble built on a 2x6 tongue-and-groove subfloor covered by 15-lb. felt and a stucco screen.

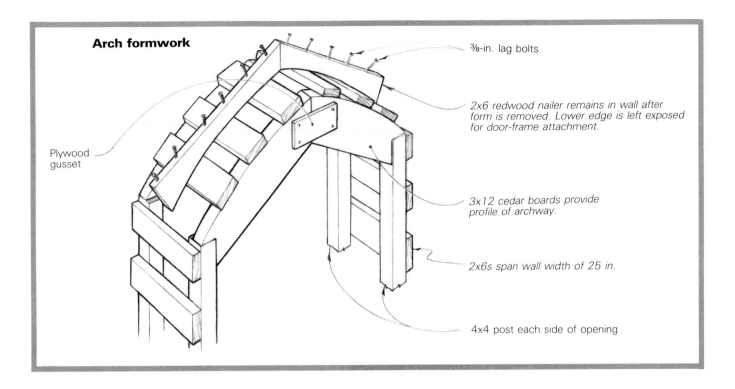

Arch formwork

⅜-in. lag bolts

2x6 redwood nailer remains in wall after form is removed. Lower edge is left exposed for door-frame attachment.

Plywood gusset

3x12 cedar boards provide profile of archway.

2x6s span wall width of 25 in.

4x4 post each side of opening

The formwork for this 5-ft.-wide arched doorway into the basement on the south side of the house was built from four old 3x12 cedar boards that were on the property. Monson used the system shown in the drawing above.

Monson built the formwork for his first arch, the 5-ft.-wide doorway into the basement on the south side of the house, out of four old 3x12 cedar boards that he had on the property. Each pair of 3x12s is held together at the top with a plywood gusset and supported by 4x4 posts. Pieces of 2x6 span the wall (25 in. thick at this point) and connect the arches; the 2x6s also support the stones that form the arch.

When the wall was built up to the foot of the arch, Monson set a redwood nailer into the arch so that he would have a place to nail the door frame and stops. He cut two pieces of 2x6 the length of each half of the arch, then cut the curve of the arch into the bottom of each one. He centered these in place on top of the forms, nailed them in and screwed ⅜-in. by ⅜-in. lag bolts into them until they looked like 2x6 porcupines. (The bolts are screwed in only far enough so that they won't come out, not enough to go through; if bolts weren't used, the nailers would be held only by the concrete, and would likely pop off.) Then he began to lay stone atop the arch form and around the nailers. The concrete worked into and between the bolts embedded the nailer firmly in the wall; using redwood ensured that the nailer would not decay in contact with the stone and concrete. When the forms were pulled, the edge of the redwood nailer could be seen on the underside of the arch.

Monson's forms for the pointed Gothic arches on the first floor of the house were much simpler (the openings were also much narrower). For these he bent 20-in.-wide, ½-in.-thick plywood to the correct arch shape and braced it with 2x4s. He built redwood nailers into these arches, too, using the method just described. He used 2x6 stock for the window and door jambs, and porcupined the sides that would be embedded in the stone with ⅜-in. by ⅜-in. lag bolts. The door sills are redwood with the same bolts extending into the masonry. The finished thresholds are locust boards Monson milled with his homemade bandsaw mill.

The Monsons worked early spring through late fall. The days were usually warm, but night-time temperatures dropped below freezing, and they had to cover the walls with old blankets to protect the curing concrete. After a freezing-cold night, Monson often woke to find the aggregate frozen, and so he invented an aggregate heater. The idea came from *One Day in the Life of Ivan Denisovich,* Solzhenitsyn's novel that describes life in a Siberian labor camp. The heater was simply a plate of ¼-in.-thick steel supported by concrete blocks on the corners with a wood fire underneath. Monson would dump four buckets of aggregate at a time onto the hot steel, and it thawed in no time.

Most of the wiring and electrical boxes are in the interior cabinets or partition walls. There are a couple of boxes right in the stone, however. The boxes are mortared right in, and flexible plastic-shielded BX cable is run up among the stones and concrete to the top of the wall, where it is run along the wood joists or hidden along the edges of upstairs cabinets and storage areas.

At the top of the walls are double 2x12 floor joists separated by 2x blocking. These are 3 ft. on center and are cantilevered out 2½ ft. past the walls. They are drilled for the top run of ½-in.-dia. rebar, then set in place with stone and concrete where they cross the walls. Monson wanted these joists to be perfectly level, so he put them in place before the walls reached their full 10-ft. height with an elaborate system of shoring and cross-bracing that made it impossible to move through the inside of the house. Then he laid stones up to the pre-leveled joists, picking and choosing the stones that would fit perfectly.

The floors and countertops

The stone floor downstairs in Toad Hall is made of rocks and pebbles laid in a mortar base. Throughout the floor there are animal patterns made of darker-colored rocks. Underneath the rock is 2x6 tongue-and-groove flooring covered with a waterproof layer of rolled 15-lb. tarred building felt. (The latter keeps the wood from drawing all the moisture out of the mortar.) The mortar is reinforced with a layer of stucco screen, which is nailed to and held above the felt with ⅜-in. furring nails.

The Monsons used a mortar mix of one shovel of cement to four shovels of sand to a half-shovel of fireclay with enough water to make it right. The mortar was laid down about 2 in. thick and screeded off, and the designs for the animal patterns were scribed into the wet surface. The pebbles were placed gently on the surface of the wet mortar within the pattern lines. The Monsons then set the pebbles in the mortar to a consistent depth by standing on a 2x12 and tapping it with a big hammer. They worked a section at a time—the 2x12 was supported by the previously finished section on one end and a 2-in.-thick board on the other so it wouldn't dip over on the edge or get pounded too deeply in any one section.

The Monsons kept a clean, sharp edge on each section of floor, making sure not to walk on it before it had cured, so it would bond well to the next section. After each pour, they sculpted around the stones on the edges and scraped away the loose mortar with a spoon and a putty knife. (They also

The double 2x12 floor joists on the west side of the stairwell rest on a 6x6 beam supported by oak log posts.

Animal designs are set with contrasting stones in the pebble floor.

kept a good supply of vinegar on hand to neutralize the alkali from the mortar on their fingers.) Once the floor was all set and dry, they wire-brushed the surface and cleaned off the dried mortar with a mix of one part muriatic acid to four parts water—a thick paste of baking soda was kept handy to neutralize the acid when it splashed on skin or clothing. Once cleaned, they hosed the floor off and let it dry. Monson put a coat of water-emulsion wax on the floors, but he doesn't feel that it did much good. For a permanent shine, he recommends mineral oil, which gives a wet look without the bright shine of a heavy-duty sealer like Varathane.

The stone countertops in the kitchen and bathroom were laid the same way but with fancier pebbles. To lay the pebbles on the vertical surface behind the stove and in the shower, Monson used ¾-in.-thick plywood, tarred felt and stucco screen furred out with the same furring nails he used on the floor. But he limited himself to working only about a foot of wall per session, the problem being that if he went too far at one time, the whole thing would slump right off the wall. He tried to leave the cut-off point jagged to help prevent the slump, and so the new work would blend in without obvious breaks where one session ended and the next began. It helped to stop on the edge of a pattern, if possible, so that any break in the work would simply emphasize the pattern. Pebbles and mortar on these walls measure about 1½ in. thick. When he does a shower, Monson first puts up three layers of hot tar and building felt before applying the mortar and pebbles—the tar has to be hot enough to melt through the felt to create a strong, waterproof surface.

Pebblework is everywhere in the Monsons' house. The kitchen countertops and all the floors on the first level are pebbled, as are the bathroom vanity and backsplash.

The roof

The roof of Toad Hall is supported by 40-in.-high side walls framed with 2x6s on the cantilevered floor joists. Monson fabricated his own truss system. He used double 2x12s with a collar tie toward the peak; the two 2x12s are bolted to opposite sides of the 2x12 tie and also to two sides of a short 2x12 gusset at the peak. Once the rafter ends were up on the framed walls, he lifted the trusses into place with a winch.

With the rafter trusses in place, Monson engineered the hexagonally shaped purlin system. He planned to use random widths of 2x pine for the roof decking instead of plywood and wanted to give the huge trusses the shear strength they needed. He designed a system of alternating intersecting diagonal purlins made of 4x4s nailed to the tops of the trusses. These run from one end of the roof peak to the eaves on the opposite end of the roof and are spaced 5 ft. apart. The intersection of these diagonals creates a pattern of diamond shapes over the entire roof area and makes a very strong unit. The trusses, which are on 6-ft. centers, create hexagonal openings where they intersect the diamonds of the purlins—as Monson nailed down the purlins and saw these hex shapes emerging, he hit on his idea for the hexagonal skylights. At one point he was going to put five skylights on each side of the roof, but he decided that three along the bottom on each side would give him all the light he needed. The skylights ended up being 6 ft. wide (the spacing of the trusses) and 5½ ft. tall (the distance from the top to the bottom of one of the purlin diamonds). With the purlins in place, Monson started on the pine decking. The 2x boards he used vary in width from 6 in. to 14 in. and are nailed down parallel to the trusses. On top of the decking he put down 1½ in. of rigid foam insulation covered by ⅝-in.-thick plywood, tarred roofing felt, and, finally, the cedar shingles.

Finishing up

As in most owner-built homes, the Monsons still have things to complete, but the house is already full of attractive, hand-crafted details such as inlaid doors and custom wood latches. The 18-ft.-long cabinet that lines the upstairs stair opening has a beautiful counter covered with strips of assorted fruitwoods that Monson milled in his woodshop. Around the edge is a strip of red narra wood, the same wood that frames the blue tile at the tops of the downstairs walls.

The hot-water system for the house runs from the spring by gravity through the coils in the downstairs woodstove and into an 80-gal. holding tank under the house. There are also coils in the same system that run under the wood in the bathroom pebble floor for radiant heat and an exposed run of coils arranged like a sort of radiator in the bathroom for additional heat. Monson has several dozen valves downstairs so that he can run the woodstove-heated water through all sorts of combinations of routes; for example, he can shut off the radiator and/or the radiant heat. A small circulating pump in the basement moves the heated water through the system. There is no relief valve, but there is a weak tee in the copper pipe that pops open when the electricity goes out and the circulating pump shuts down. It's not the kind of system I would recommend, but it seems to fit right in with the overall spirit of the house.

The rafter trusses rest on the top of 40-in.-tall 2x6 walls. Photo by Jim Santana.

Roof purlin and truss system

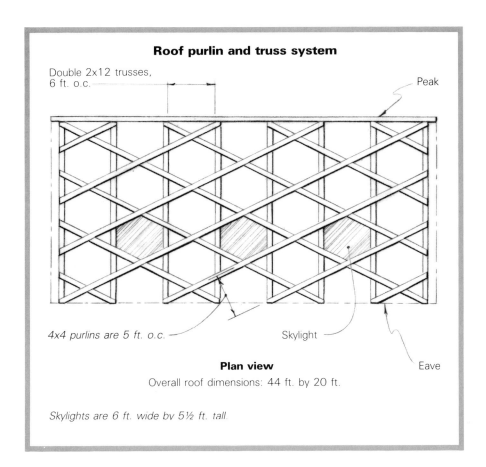

Double 2x12 trusses, 6 ft. o.c.

Peak

4x4 purlins are 5 ft. o.c.

Skylight

Eave

Plan view

Overall roof dimensions: 44 ft. by 20 ft.

Skylights are 6 ft. wide by 5½ ft. tall.

Skylights are formed by the intersection of roof trusses and purlins.

An Introduction to Log Houses

CHAPTER 5

Log houses, like stone houses and earth houses, were originally built because the material was handy and logs made good, solid homes. But while earth and stone became closely associated with the drawbacks in their methods and their link to a poorer past, logs somehow rose above their limitations. Until recently most people were convinced that an earth house was going to wash away. Most people thought that stone houses were always damp, cold and mossy. But only the people who are actually building (or living in) log houses worry about the chinking popping out or about wood preservatives, about the walls shrinking, uneven floors and difficult housekeeping. The rest of the population seems to accept log building as just another type of wood-house construction, albeit one that is inseparable from the folklore. Kerosene lamps, handmade wooden furniture, animal-skin rugs and hangings, the rifle over the mantle, the big stone fireplace—these romantic images are embedded in our collective imagination.

Log building has its own unique set of challenges and problems. As with any building method, the challenges are more easily met if the builder is experienced; difficulties arise when the first-time builder has not done adequate homework or is unable to sort through the options, all of which seem equally valid. There are several areas that demand the close attention of log builders, and within those areas are endless opinions on the best way to do the job. As with rammed earth or stone building, there is no one best way—instead the solutions to the common problems of log building vary from region to region and from builder to builder. Over and above the normal problems of building a house,

log builders have to pay particular attention to selecting the tree species and preparing the logs; stacking the logs into walls; chinking, caulking and finishing; allowing for shrinkage and settling; cutting in doors and windows; and planning the roofing, foundation and mechanical systems—all of which I'll discuss in a moment.

Problems aside, if your dream is to build a log home, there are lots of ways to do it. The simplest way is to order a kit from a catalog and hire a contractor to build it for you. My neighbor Rick Schaad, who is a log-house general contractor, builds his homes from a standard kit plan that he customizes for each of his clients. (Kits are usually based on milled logs, assembled in factories, sold by distributors and built by franchise builders from a few standard plans. Most kits also include doors and windows and the roof system. The 1985 *Log Homes Annual* lists about 125 companies in North America that manufacture log-home kits or packages.) Schaad built his own house, which was the first log house he ever built, from a kit (photo, facing page). He had planned to build the house on his five-acre site in the pine trees using traditional hand-hewn methods, but several factors caused him to change his mind. To use his own trees would have wiped out his small forest. To cut the trees, haul them, peel them and let them cure was going to take an extra season and maybe two. He was making his living building houses, and felt he couldn't take himself out of the business for very long. The more he thought about it, the better the kits sounded—dry logs all peeled and milled, notches and splines cut, numbered systems that were quickly assembled. He didn't particularly like the idea of living in a house of someone else's design, but figured he could personalize it during construction and finishing, which he did.

By contrast, the hardest and least expensive way to build a log home is to buy a piece of property with suitable trees, harvest the necessary trees with a chainsaw, transport them to the building site and build the house from scratch. This is the way Sue and Lance Joyner of Orcas Island, Washington, built their home. They built the house in two stages—the original 28x24, one-bedroom house, and an addition that

Contemporary builders like Rick Schaad have brought log houses right into the 20th century. Seven years ago, Schaad built his 1250-sq.-ft. house from a kit. At the time, the kit cost $13,000, but he added many extras, such as a rock fireplace, porch, decks (one surrounding an in-ground pool), railings and dormers in the second floor.

includes a master bedroom and bath and a studio for stained-glass work. The entire house cost about $25,000 to build, and the Joyners did nearly all the work themselves.

In between these two extremes there are a number of other log-building options. You can purchase a custom design from a kit company and then put the house up yourself. (But you should keep in mind that there is more involved than just stacking the pieces on top of each other like Lincoln Logs.) Another alternative is to buy the logs and have them delivered, then take the house from there. Or you can buy logs that have been milled to a uniform size, which are barkless, clean and totally without taper. You can even have the logs planed flat on one or more sides and fit with tongues and grooves so they will lock together. And you can always hire help.

If you live in an area of the country with a log-building tradition, such as the Northeast, the Appalachians, the Great Lakes region and most of the West, you can also choose to work with a hand-hewn log building company. These com-

panies hand-build log houses out of unmilled, hand-peeled logs that are fitted together, numbered, dismantled and shipped to your site along with a roof system, windows and doors. The houses tend to be one-of-a-kind designs and are not kits. The log work on the house in Chapter 7 was done by hand-hewn log builder Steve Kenady of Eastsound, Washington; the photos on p. 2 and at the top of the facing page show more of his work. The house discussed in Chapter 8 was also built by a hand-hewn log builder: John Lee of Somerset, California. Another example of this type of construction is the home built by Joel and Marie Smith, which is shown on the facing page. Their house is of lodgepole pine, and the design is far from the typical log ''box.'' To brighten the interior, the Smiths included dormers in the gallery on the second floor and windows in the gable ends.

From this sampling of houses, it's easy to see the scope and variations possible in log building. Yet the considerations involved in this type of construction are the same for all log builders.

Traditionally, log houses were built with nothing but an ax, such as the example above from the Great Smokey Mountain National Park in Tennessee. Photo by Brent Harrington.

The interior of the Knapps' home is warm and inviting.

The log-house addition built by Tish and Gene Knapp and Steve Kenady on Orcas Island, Washington, is an unusual attachment to their 1890 vintage farmhouse. This view of the west side of the house shows the stonework that covers the cement-block basement walls.

This hand-hewn log house, built by Joel and Marie Smith of Ellensburg, Washington, is made of 12-in.-dia. lodgepole-pine logs. The variety of roof lines breaks up the boxy look that log homes sometimes have. The dormers provide light in the second-floor gallery, and the windows in the gable ends brighten the living room. Photo by Dean Rogers.

Selecting and preparing the logs

Logs are wonderful things all peeled and shiny and ready to make into a house. They have incredible strength and resilience and will last a long time if protected and used properly. The species you select depends mostly on personal preference and what is available in your area, although some log builders, such as Lori Bridgwater in Chapter 6, have their logs shipped from far away. Throughout the United States and Canada, pine and fir (especially lodgepole pine, Douglas fir and white fir) are favored species, but cedar, spruce, hemlock and oak are also widely used. It's a good idea to talk to log builders in your area, because some species, such as poplar, that are shunned in certain sections of the country are used regularly in others. The size of the logs you use is also a matter of choice. Large logs look more impressive in a building, but you need special equipment to handle them and hoist them into place. Small logs require more handling (there are more of them to peel and notch), but are easier to move. Beware of logs that are too small, however, as these will require endless notching and result in a very thin wall. The logs used by the builders in this book vary from 8-in.-dia. milled logs to the 2-ft.-dia. logs shown in the photo at left.

Preparing the logs involves cutting them at an optimal time of year, culling any diseased or rotten sections, and peeling off the bark. There is much debate about the best time to cut and peel trees for logs. Most builders know the ideal conditions for gathering logs and peeling them, but the best time for harvesting is not always the most convenient time for getting the logs out of the woods. In spring and summer, roads are passable and the ground is firm. When the sap is running at this time of year, the cells of the tree are packed with moisture, which makes peeling the logs a simple matter—the bark comes free in big strips and the logs are quickly made clean and white. Unfortunately, this same abundance of moisture also makes the logs heavier and more subject to checking and twisting as they dry. On the other hand, logs felled in late fall and winter are drier than at other seasons and cure better. And because they don't have the slippery wet coating of a summer log, they are less likely to pick up surface mold and fungi. Yet in winter, roads are impassable, fingers freeze on the chainsaw, and tractors and pickups bog down in the mud. In addition, the logs are more difficult to peel and often are marked with pieces of brown bark, as shown in the photo at left. This isn't really a problem, though, and some people prefer this look.

In a log winery under construction by John Lee (top left), the logs average 2 ft. in diameter, giving a wall that is nearly 8 ft. tall and only four logs high. Lee uses short log sections around windows and doorways to cut down on waste, rather than building a solid log box like some builders do and cutting out the doors and windows later.

Pieces of bark left on the logs of the home shown at left result in a mottled look, and are common in trees harvested in fall and winter. In those seasons the bark doesn't peel as cleanly as it does when the sap is running in the spring.

You should reject badly bent or twisted logs, although you can always find someone who has used an inferior species of tree and unbelievably knotty, twisted trunks and made them work. The idea is to load the equation in your favor wherever possible, and in log building this means to use the best materials. Logs are easily peeled when green, most easily when the tree is first felled. But because the bark of a tree protects it during skidding out of the forest and delivery to the site, most builders don't peel off the bark until the logs are on site and ready to go. Tools used for peeling vary from builder to builder—drawknives and peeling spuds (which look like giant chisels) are generally favored, but I know one log builder, John Herseth, who peeled most of his logs with a sharpened turning spade. With a drawknife you can sit on the log and pull the tool toward you. With a spud or sharpened spade, you push away as you would with a chisel. (The long handle of the spade allowed Herseth to stand on the log.) The spud can be used to pare down high spots on the log as well as to remove the bark. Many builders crib the logs up off the ground and stand beside the logs as they work with spuds and drawknives.

To use a drawknife to peel a log, you pull the tool toward you (above left). The action with a peeling spud is just the opposite—you push away from you (above right).

A *V*-notch sawn in the bottom of each log creates the snug log-to-log fit characteristic of Scandinavian-style building. Most builders stuff the notch with fiberglass insulation.

Typical kit-house connecting system

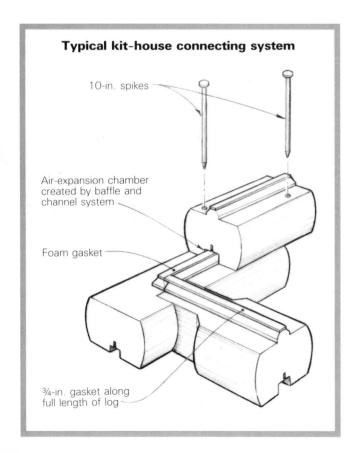

10-in. spikes

Air-expansion chamber created by baffle and channel system

Foam gasket

¾-in. gasket along full length of log

Arranging the logs in the wall

To make walls, the logs have to be stacked in such a way that they won't fall down or pop out. In the old days in northern Europe, where log building traces its roots, walls were held together with overlapping and interlocking corner notches. Some systems involved a few cuts on each log end, others involved complicated layouts and sophisticated joinery. Today the assembly of logs into walls can be as straightforward as stacking logs that have been milled flat on the top and bottom and spiking them together, with alternate logs running long at the corner. Or the log work can be as complex as the Scandinavian, or chinkless, style of building, where logs are shaped and cut to lock at the corners, and cut along their entire length on the bottom so that each will fit snugly onto the one below. The Scandinavian style of building eliminates the gaps between logs, so no filler (chinking) is required. In manufactured kit houses, there are a variety of connecting systems available, all based on taper-free logs that have been run through a milling machine. The system used by Rick Schaad (whose house is shown on p. 48) is fairly typical: tongues, grooves and a series of plastic splines and foam gaskets cut down on air infiltration between logs. (This is the Real Log Homes Interlock System, shown in the drawing at left, which is manufactured by Real Log Homes, Hartland, Vermont.)

Early on in log building, two basic styles developed—round-log and square-log construction. Each style had its own system of notching and chinking and its own set of tools. Today round-log construction is generally found in the Northwest (including western Canada), and square-log construction is found mostly in the East and the Appalachian Mountains. The preferred notch in round-log construction is the round, or saddle, notch—as much as half the log diameter is cut away from the bottom of each log, which then nestles on the log below. For square-log builders, the dovetail notch or the *V*-notch is used. There are many variations of these notches, and occasionally builders combine round and square notches, or even square the ends of round logs and then dovetail them.

The whole business of notches and which one to use where can be pretty confusing to novice builders, some of whom are reluctant to take the time to learn the skills involved and to cut all those log ends. For them, the crib corner shown in the drawing on the facing page is an easy solution. The house in Chapter 6, built by Lori Bridgwater, uses this construction; John Herseth also used it to build the house shown in the photo on the facing page. But unlike Bridgwater, Herseth used round logs, which didn't butt together like squared logs, so he had to use a lot of extra insulation in the openings and extra chinking and caulking.

Whichever system you choose, let the logs run well past the corners of the building to retard rot and decay. Because the end grain of a log absorbs water more readily than the rest of the log, it may start to deteriorate and rot back to the joint. The longer the log projects beyond the joint, the more protection the joint has. In traditional log houses, which were held together by the joints alone (no spikes or steel strapping), early builders added years to the lives of their houses by developing joints that would shed water and resist rot.

Types of corner joints

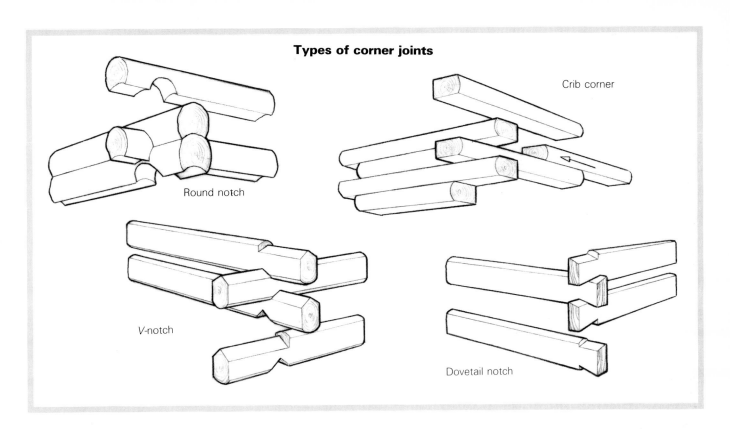

Round notch

Crib corner

V-notch

Dovetail notch

The log home belonging to John and Betty Herseth, of Seattle, Washington, has lapped and butted corners spiked together with ½-in.-dia. rebar. This system creates large holes to be filled where the round logs meet, but Herseth thought that it would be easier to master insulating and chinking than to scribe and cut all the notches for the 1,000-sq.-ft., log-enclosed first floor of the house.

Cutting the notches Like most skills involving movement in several directions at once, concentration and careful measurement, cutting a notch is more easily demonstrated than explained through words. The basic saddle notch is not that hard to cut, although it can take hundreds of notches before a first-time builder masters the process. To give you an idea of the technique, I'll go through the steps in cutting a saddle notch here.

To start, imagine a simple, rectangular log house with two short walls and two long walls. The first logs, the long ones in this example, must be flattened to rest on the foundation or notched to rest on top of foundation piers. The short logs will then be scribed and notched on both ends where they rest on the long logs. All subsequent logs will be scribed to the logs below them and notched. To transfer the contour of the lower log to the surface of the upper log so that a snugly fitting notch will result, a log scribe is used. This is just a large version of a compass, available in many designs. The log scribe has two adjustable legs, one of which holds a pencil, pen or lumber crayon; the other has a solid metal point. (Some scribes have a level attached to the handle so that the legs can be kept strictly perpendicular during use, which is necessary for accurate scribing.) The legs are locked a given distance apart with a screw.

The log to be scribed must be secured with log dogs to the log beneath it. The logs must be aligned vertically and horizontally, as shown in the drawing below. To align them vertically, you can mark a vertical line down the center of each log end with a large carpenter's level before scribing, or eyeball alignment against sight lines. Horizontal alignment is important because trees are much fatter at the butt end

Here, Neal Nygard uses a log scribe to mark the top floor joist of a house.

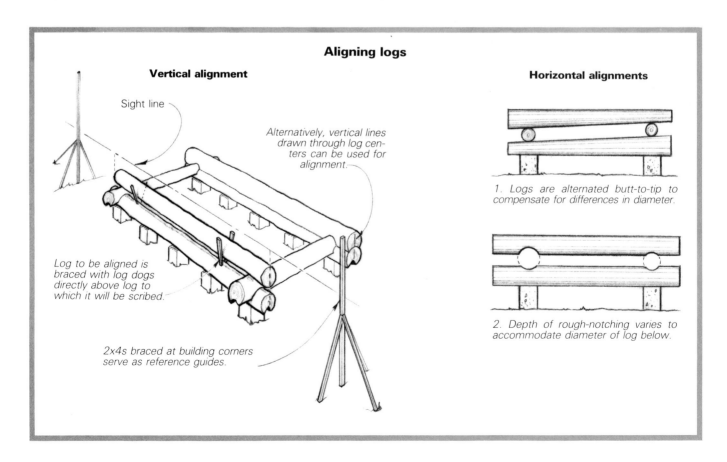

Aligning logs

Vertical alignment

Sight line

Alternatively, vertical lines drawn through log centers can be used for alignment.

Log to be aligned is braced with log dogs directly above log to which it will be scribed.

2x4s braced at building corners serve as reference guides.

Horizontal alignments

1. Logs are alternated butt-to-tip to compensate for differences in diameter.

2. Depth of rough-notching varies to accommodate diameter of log below.

than at the tip—usually the first section, cut from the butt, tapers radically and can't be used. A good tree for a log home is one that grows tall and true like a telephone pole with little taper. But even these trees will still have a butt end that's wider than the tip. One way to compensate for this is to reverse the logs end-for-end as you stack them: butt over tip, tip over butt, and so on up the wall. Any problems in horizontal alignment can then usually be solved by cutting a deeper corner notch if there's a bulge in one area, or by using a thinner log in the next course. It's important to always think one course ahead.

To scribe the notch, adjust the log scribe to half the diameter of the lower log. (Experienced log builders have learned to fine-tune this scribing business. Some set the scribe for a little less and cut a rough notch first, test the fit and then rescribe for the final cut. Others adjust the scribe to the maximum gap between the logs, then add ½ in. Some keep test-fitting as they cut the notches, working gradually to the scribed lines.) Lock the points of the scribe and hold it perfectly perpendicular to the ground, one leg over the other. With the metal point of the scribe on the side of the bottom log and the pencil point on the bottom of the top log, move the scribe up the side of the bottom log, which will record all the bumps and lumps of the bottom log onto the top log. To complete the arc, move to the opposite side of the bottom log and repeat the process. Do the same thing to the opposite side of the top log, so that there are two cut lines. Go to the other end of the log and scribe the notch there, then roll the log toward the inside of the house with the marked half of the log up, wedge it in place and chainsaw the notches.

Cutting the notch with a chainsaw is best done in stages. Most builders make a rough notch first. Some cut a line down the middle of the notch, taking care not to cut into the scribed line, and then angle the second and third cuts to meet at the end of the first cut. They then shape right up to the scribed line with a mallet and chisel (or the chainsaw). Other builders run a series of cuts parallel to the first one and shy of the scribed line, then knock them all out with an ax or a large chisel. The secret to cutting a good notch is to cut right up to the scribed line but not into it. If you do that and the notch doesn't fit, then the scriber has made a mistake. If the line has been cut away, you can assume the cutter made the error. If you find that you are both scriber and cutter, be careful on both parts of the job, because you won't have anybody else to blame if your notches don't fit. You'll have to rescribe and recut, or throw away the log.

After you've cut the notch, the top log should fit tightly against the log beneath it. Builders who want logs that fit snugly along their lengths so that no chinking is required (the Scandinavian system) scribe the length of the log at the same time they scribe the end notches. Then they chainsaw a V-notch along the entire length. The same three basic cuts can be used for this as for the notch—one cut down the middle of the length of the log, and an angled cut on each side of the first cut. The V-shaped notch is then rounded and shaped with either chainsaw or ax. Most builders who use this system scribe and cut the log length a little deeper than necessary and then fill it with fiberglass insulation as they stack the logs.

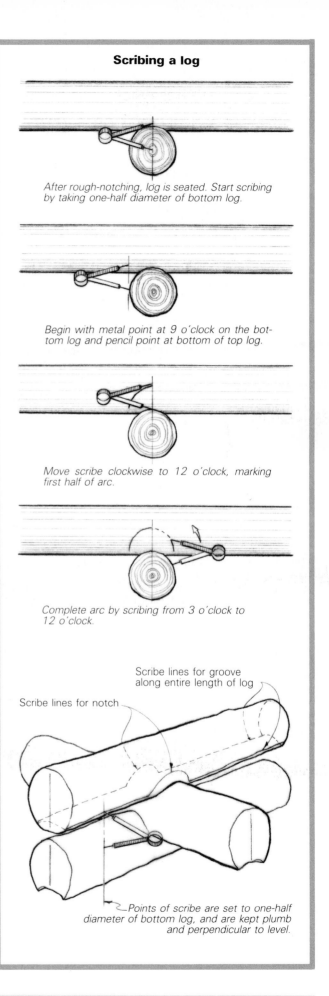

Scribing a log

After rough-notching, log is seated. Start scribing by taking one-half diameter of bottom log.

Begin with metal point at 9 o'clock on the bottom log and pencil point at bottom of top log.

Move scribe clockwise to 12 o'clock, marking first half of arc.

Complete arc by scribing from 3 o'clock to 12 o'clock.

Scribe lines for groove along entire length of log

Scribe lines for notch

Points of scribe are set to one-half diameter of bottom log, and are kept plumb and perpendicular to level.

Log-builder Steve Kenady agreed to incorporate a dovetail corner into Gene and Tish Knapp's house as long as it was covered by the greenhouse that connects the log addition to the old farmhouse (top). To seal out drafts and moisture, the gaps between the logs must be filled with some type of chinking. The chinking between the dovetailed logs shown here (bottom) is white cement mortar on wire.

Instead of building his walls in the chinkless style, Steve Kenady, whose work is pictured on pp. 2 and 51 and in Chapter 7, cut a kerf down the length of the top and bottom of all his logs and inserted plywood splines. The idea was that the spline would provide a good backup for the chinking and cut air infiltration even if the chinking should fall out. Log builders generally agree that any cut down the length of a log prevents excessive checking and twisting, but there is disagreement on how deep the kerf should be or if it weakens the log. The Joyners (mentioned on p. 49) planned to use the Kenady spline system until Lance Joyner was involved in modifying an early Kenady house. When he cut into one log wall, he found that moisture had gathered around the spline in the kerf on top of each log. The Joyners decided to leave out the spline, but still kerfed the bottom of each log to prevent checking. They then put oakum between the logs where they touched and used cement mortar to fill the gap on the outside.

The most complicated notch work is probably the dovetail used by square-log builders. There are several reasons why it's a particularly good joint. For one, you don't cut as much of the log away to make the notch. (A round notch wastes almost half the log.) In addition, the cutting that's necessary wastes the decay-prone sapwood while leaving the stouter heartwood. If the shoulders of the joint are angled down, the dovetail sheds water to the outside of the building. Aesthetically, dovetails are attractive. Tish Knapp liked them so much that she had her house redesigned just so she could have one dovetail corner in her round-log house, as shown at top left. Another advantage of the dovetail notch is that as the wood shrinks and the joint loosens, it can be driven tightly back together with a large hammer. The only drawback is that dovetails are complicated to lay out and cut. With square logs and dovetails, you're also committed to a heavy program of chinking, because the notch doesn't take up the gap between logs the way a saddle notch does. Some books and articles on cutting dovetail notches are listed in the bibliography.

Chinking, caulking and finishing

Keeping out the weather is a major concern in log building. If you don't use the Scandinavian style of building, how do you stack a bunch of uneven logs atop each other without ending up with gaping cracks between them? Any gap between the logs will allow warm house air to escape and cold outside air to intrude. To block this air exchange, some method of filling, or chinking, the cracks must be used. Our forefathers packed the gaps in their walls with mud; for really large cracks, they first inserted wooden shims and then filled in around them with mud. More recently, log builders have gone to more weatherproof, cement-based mortars to fill the inter-log spaces. Every filler—whether it consists of moss, sticks and mud, mortar, or a new miracle plastic—is still called chinking. Narrow cracks around doors and windows, checks in the logs, and cracks in the chinking are filled with the same caulking compounds used in conventional construction. There are lots of solutions to the problems of weatherproofing log walls, and several will be discussed in the next three chapters.

Not only does chinking keep out the wind, it also prevents moisture, which will rot the logs, from settling between them. But chinking that loosens will trap moisture, so the chinking must be maintained. The problem is that as the wood shrinks and swells in response to humidity in the air, the movement is sufficient to crack the cement-based chinking, which expands and contracts at a different rate than the wood. And so the chinking works free. Some builders insert a wire base between the logs to hold the chinking, others use mortar formulas that have some flexibility. The Gilbertsons (Chapter 7), for example, ran single-strand wire diagonally between nails and mortared over this. To spend more time notching the logs for a chinkless fit or to depend on chinking is a decision every log builder must make. The former is more work initially, but maintenance on the latter must be done for the life of the building.

The same log movement that can loosen the chinking is also hard on the caulking. Some caulking is like old-fashioned window putty and it dries hard and inflexible. Silicone-based caulking dries but never hardens, remaining flexible. It's best to use a flexible caulking to seal the cracks around doors and windows and the tops of the walls. Lori Bridgwater (Chapter 6) ended up using caulking to seal the cracks that opened between her flat-sided logs. It's a good idea to caulk the splits that develop in the tops of the logs, to keep moisture from getting into them.

Wood is a superb building material, and under ideal conditions will last indefinitely, but most buildings exist under less than ideal conditions. The exposed logs of a log house are susceptible to weathering and attack from moisture and wood-digesting fungi. Logs are also the natural homes of termites, carpenter ants and beetles of various names. Some species of trees resist decay better than others and these are used where possible. (Redwood and cedar are species that have natural decay-resistance, but they are not generally available.) Because moisture is a key ingredient in the attack of fungi, and wood with a moisture content below 20% does not support fungal life, the smart log builder builds his or her house from peeled logs on a good foundation that will keep the logs off the ground. A good roof overhang will protect the walls from rain and snow, and gutters take care of splashback.

In addition, sealers sprayed or painted directly on the outside of the logs help keep moisture out. All sorts of products have been used over the years, from diesel and motor oil to expensive, imported penetrating sealers that include fungicides, insecticides and ultraviolet-reflecting pigments. Basically there are two types of coatings: ones that create a paint-like film on the surface of the log (these may be clear or pigmented), and ones that penetrate the wood and harden within the pores. Which type you use is again up to you.

Generally, a dry log will accept a coating better than a green one. Pentachlorophenol (penta) is widely used as a wood preservative when diluted to a 5% solution, but use penta with care. Both the Environmental Protection Agency and the Center for Disease Control have issued warnings about its dangers, and the U.S. Forest Products Laboratory cautions that it should never be used *inside* a log house. Many log builders use solutions of linseed oil, penta and

Keeping joints dry

A chinkless log wall, either V-notched (left) or laterally grooved (right), sheds moisture.

On a chinked wall, a concave finish provides good drainage (left). Avoid thin feather edges (right), which will eventually break.

paint thinner as a sealer. Some substitute diesel oil for the thinner, but diesel darkens logs. A product called log oil, made from a varnish base with linseed oil and paint thinner, is also popular—I've heard of some log oils that never dry and stay tacky forever, though, so you should test the brand you buy before you use it. Some builders dilute their log oil fifty-fifty with mineral spirits and spray it on their buildings.

Whatever sealer you use on the outside of the walls will have to be reapplied every two or three years. A lot depends on weather conditions, the way the sealer is put on, and the exposure of the walls to sunlight and rain. For interior finishes, log builders use the same sealers and coatings as conventional builders who are covering bare wood for a natural look. These include varnishes, clear acrylics, penetrating oils and transparent stains either sprayed or brushed on.

Shrinkage and settling

You really hear about shrinkage and settling from first-time log builders who just didn't realize the problems that are created by logs that are allowed to cure in the walls of a house. If you don't plan for this movement, you'll wind up with wavy floors, doors that won't open, windows that break, cabinets that pull away from walls and divider walls that buckle. It's a shame that these problems still develop, since log builders over the years have worked out so many ways to prevent them.

Shrinkage happens as the wood dries, and the greener the wood, the more pronounced the problem. In other words, if your log wall is 10 logs high and each log ultimately shrinks ½ in., the top of your wall will be 5 in. lower than it was when you laid the last log in place. Settling takes place as the weight of each log bears down on the one below it, flattening the crown and tightening up the fit. To cope with both these eventualities, you can use dry logs. You can also let the house sit and settle once the logs are up before proceeding with the rest of construction. Or you can construct special systems throughout the building that will take up the slack. Most log builders allow for settling as they build, leaving a 2-in. to 4-in. gap at the top of all windows and doors, and allowing for movement in interior stud walls. The trouble comes with support posts in the middle of a building that either try to punch through the floor at their bases or, where there is a second floor or loft, the floor above. (Joel and Marie Smith, whose house is shown on p. 51, have screw jacks at the base of every post, which are adjusted to match the settling of the walls. The same system is used by the Densmores, whose house is covered in Chapter 8, and you can see it in the photo on p. 92.)

The unplanned settling of cabinets and counters is especially disconcerting. Almost all log builders have stories of carefully scribed and scalloped cabinets or walls that fit beautifully the first couple of months, but then later look like they were cut to fit a different house. Contractor Rick Schaad gets around this problem by making his cabinet units independent of the log walls. In a recent project he had a tall cabinet at each end of a wall—one for the refrigerator, the other for the oven and microwave. A beam run along the top of the two tall cabinets was connected to the bottom cabinets with a sheet of plywood, which served as underlayment for the tile—the whole thing was freestanding.

For his interior divider walls, Schaad uses a system similar to that described in Chapter 6. He attaches the last stud to the log wall with a lag screw; a vertical slot cut in the stud allows the log wall to move without stressing the screw or stud wall. He leaves a space at the top of the wall, and doesn't apply molding until the house has settled. Some builders top-nail the molding to a nailer in the ceiling but don't nail it to the divider wall. This way the molding can slide down over the wall as the logs settle. Molding over the gap at windows and doors can be handled the same way. If it is top-nailed to the log header, it will float over the window or door frame and will not buckle or crack as the logs settle.

Cutting in doors and windows

A log box locked at the corners is a very strong structural unit, but when you cut into the walls, you begin to lose some structural integrity. Therefore, some system must be used to retie the logs and keep them in line where an opening has been cut. Spikes, various sorts of splines and short log systems are all used, and you'll see several different applications in the following three chapters.

Cutting in doors and windows after the walls are up has some advantages over building in the doors and windows as you go. The log work is generally faster because you don't have to deal with tying the short logs in the openings together, and you have the luxury of deciding where and how big the windows will be after you've have a chance to sit in various parts of the house. On the other hand, it's really a waste of good logs to cut into your nice, straight, 40-ft.-long logs when you could have used short ones in the first place—butt logs and scraps that would otherwise go to the woodpile can often be used for short sections by the builder who puts in door and window openings as the building progresses. Here again, what method you use is a matter of personal preference.

Here, window and door openings are grooved out. A 2x4 brace will be nailed in to hold the opening square and the short logs in place.

Roofing, foundation and mechanical systems

The roofs of log houses can be just the same as for any other house, but there are some characteristics that mark the traditional style. Many builders opt for an exposed cathedral ceiling to show off massive ridgepoles and purlins. Fancy log trusswork is common in handcrafted log homes. The standard pattern is to have a purlin or two on each side of the ridgepole spaced between it and the top log of the wall, as the Gilbertsons did (Chapter 7). They also used the top wall log as a purlin and ran the decking perpendicular to the eaves without rafters. Other builders use purlins and rafters with decking parallel to the eaves, as Lori Bridgwater did (Chapter 6). Some builders run the logs all the way to the top of the gable ends, and some frame the gables with standard framing and insulation. Most prefer to keep ridgepoles, purlins and rafter ends under an overhang to keep them out of the weather.

A house made of logs needs a heavy foundation, but like other aspects of log construction, there's no one right way to do the job. I've seen foundations of log houses made of rock, wood or concrete piers, and poured concrete; some builders use massive footings and stem walls, and others use the size footings and walls you'd expect to see under a stud-frame house. But a point on which all log builders seem to agree is that the foundation must support the logs at least 1 ft. off the ground to keep them dry and to let air under the house.

A reinforced-concrete perimeter foundation provides the strongest support for most buildings. If a portion of the footing is washed out or settles, the concrete beam will span the gap. A house on a pier system depends on the integrity of every pier. If one settles or is washed out, the entire house sags at that point, but if the piers are done carefully and all dug down to stable soil, there should be no problems. The advantage of piers is that they save money and work. The site does not have to be level, so you save money on tractor work. You save money on digging the trench all around the house, and you especially save money on the concrete and steel needed to fill that trench.

Some people who build the concrete foundation with stem walls are bothered by the contrast of smooth concrete with natural log walls. A good covering in this case is a facade of stone. It creates a more natural look and adds a feeling of strength. Rick Schaad's house (photo, p. 48) was done this way, as was the house featured in Chapter 8.

Log houses are similar to earth and stone houses in that it is not too difficult to get the electrical and plumbing runs into the walls if they are done at the time of construction. But it is very difficult to add circuits after construction is completed. Holes can be drilled through the logs as they are laid up and electrical boxes can be cut into place. Non-log interior divider walls, cabinets and closets are all good places to hide wiring and plumbing. In a log house, a common solution is to run the wiring around the bottom of the sill log and to run plumbing in divider walls. Some builders mount their electrical outlets in the floor and wire them from under the house. Some create a sculpture out of the exposed pipes.

The fancy trusswork on this unpeeled-cedar-log house is more decorative than functional. A steel rod 1 in. in diameter runs from the horizontal log to the ridgepole above. Photo by Charles Miller.

The House at Tigard

CHAPTER 6

The log house that Lori Bridgwater built for herself and her family on the wooded outskirts of Tigard, Oregon, is an impressive first house. All told, it offers nearly 4,000 sq. ft. of attractively arranged living space. The house is built of stacked and spiked 8-in.-dia. logs milled on three sides and uses log beams and posts throughout, but it is as far from a log cabin as you can get and still be using the same materials.

When the family bought the Oregon property, there was already an old stud-framed house on it, complete with half-round logs nailed on as siding. The house had started out as a summer cabin, but then the previous owners had added on a porch, changed the porch to a room, added a second wing and eventually added a second story. Originally, Bridgwater had planned to remodel the existing house, but as she looked at it more closely, with its sagging floors and fake-log walls, she decided to tear it down and start from scratch. She wanted to keep the old fireplace and stonework in the living room, however, and to have, as the existing house had, a fireplace in the kitchen, so she saved the main fireplace, the footings for the kitchen fireplace, and the foundation for the living room and kitchen (which accounts for the several level changes in the house). She also kept in place the beautiful floor joists made out of old 10x10 barn beams.

Bridgwater didn't know much about log building when she decided to build their house, but she did know something about building. She had been a medical secretary when she and her husband and two children got involved in doing the finish work on their summer house on the Oregon coast. It was then that she discovered she enjoyed swinging a ham-

mer far more than she ever had typing medical reports, and so she enrolled in a construction program at Portland Community College. Midway through her courses she decided she would rather be a contractor than a worker, and as soon as she was out of school she landed her first contracting assignment in the Portland area. Her first big project, however, would be the family home in Tigard.

When the Bridgwaters decided on a log home, they agreed they didn't want a kit house, but they did want logs that were flat on the inside. For a house the size they were considering, they would have to buy milled logs if they didn't want to spend the rest of their lives shaping and fitting logs. (As it was, Bridgwater had plenty of log work to do just notching and scribing where the main support timbers on the house came together, and notching the logs in the walls around the doors and windows.)

Bridgwater also figured that logs that were flat on three sides would be easier to lay up, although the fit between logs would not be as tight as if they had been scribed. And indeed there has been some uneven settlement and some seams have opened between logs. On the Bridgwaters' protected site this isn't really a problem. But as it is, Bridgwater is busy caulking between all the logs from the outside, a job she hoped she could avoid by insulating. Now, like many log-home owners, she has the task of maintaining the caulking as the logs shrink and swell with the seasons.

Bridgwater found a good source of pine logs at a mill in Ashton, Idaho. All the logs were bug-killed, which means they were logged after they had died on the stump. The advantage, besides the marvelous blue staining that occurs in pine trees that have been killed by beetles, is that such logs have begun to dry before they're cut, and are a lot lighter to pack around the building site and to hoist into place. In addition, the drier the log, the less the shrinkage—and the less settling. I asked Bridgwater if she was worried about bugs in her bug-killed logs, and she told me that the beetles kill the tree but don't infest the wood. I've since talked to other log builders who use bug-killed pine and they also have no doubts about the soundness of their logs.

The Bridgwaters' spacious house has nearly 4,000 sq. ft. of living area and was built by Lori Bridgwater and a small crew. Extending off the front of the house is the master wing, with the children's wing and second-floor guest quarters on the north side. The stone chimney off to the left of the photo is from the original house that was on the site.

The milled wall logs Bridgwater used were uniform and free of taper. The logs for the ridgepole, purlins, rafters and posts were unmilled but hand-peeled at the mill—not having to peel the logs saved her untold hours of hard work. In all, she used $18,000 worth of logs for the project.

The Bridgwaters planned their unique home well. The large living room with its massive stone fireplace is on a level with the front entry. Two steps down on the west side is the dining room, and two steps down on the north is the 30-ft.-long kitchen and breakfast nook with its 9-ft.-high wood-decked ceiling and 4x8 skylights at both ends. The master bedroom and bath take up one wing of the house, and the children's playroom, bedrooms and bath line the north side of the downstairs. Also on the first floor is an office, laundry room, large pantry, and an extra full bath. Upstairs is another 1,300 sq. ft. of living space, which includes guest quarters and a storage room.

The foundation work for the house was begun in August 1982 and the house was finished in August 1984. Bridgwater would probably argue with the use of the word "finished," for she still has lots of little details to take care of, and wishes she and her family had not moved in until the house was really done.

The living room of the Bridgwater house has as its focal point a large fireplace and hearth, which was kept intact from the pre-existing house on the property.

The dining room is two steps down from the living room on the west side of the house. A door on the north leads into the kitchen and a sliding door leads out to a small patio.

The master bedroom has nearly 400 sq. ft. of space. Four-inch-wide cove molding at the top of the walls covers the space left for settling.

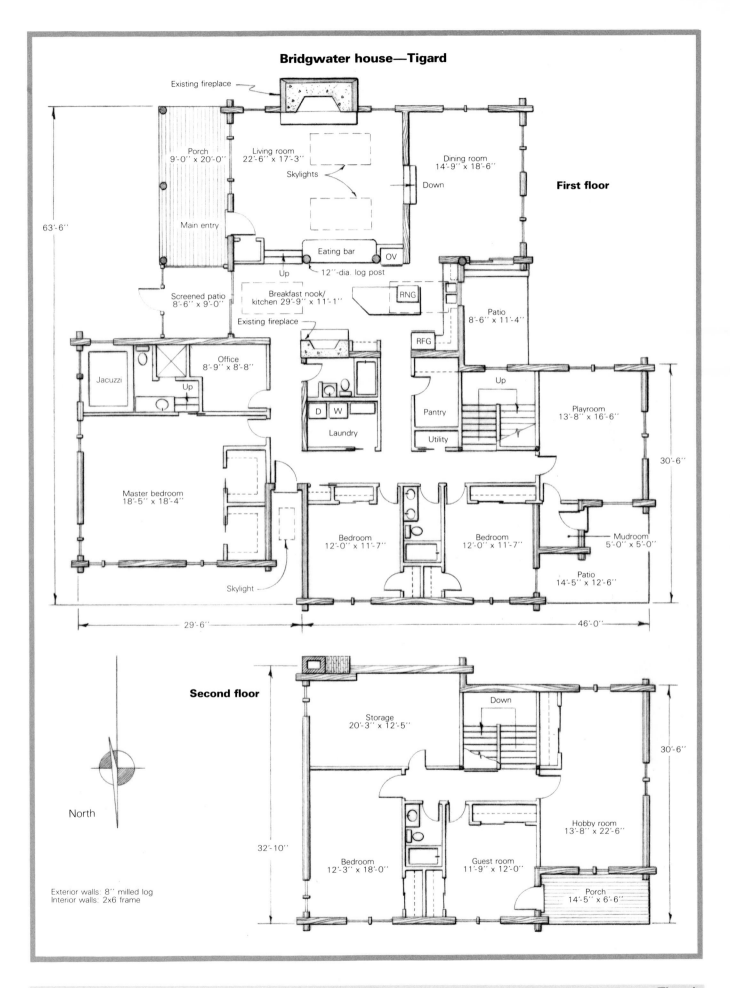

Bridgwater house—Tigard

First floor

Existing fireplace

Porch
9'-0'' x 20'-0''

Living room
22'-6'' x 17'-3''

Skylights

Dining room
14'-9'' x 18'-6''

Down

Main entry

63'-6''

Eating bar

OV

Up

12''-dia. log post

Screened patio
8'-6'' x 9'-0''

Breakfast nook/
kitchen 29'-9'' x 11'-1''

Existing fireplace

RNG

Patio
8'-6'' x 11'-4''

Jacuzzi

Office
8'-9'' x 8'-8''

Up

RFG

Up

Playroom
13'-8'' x 16'-6''

Pantry

D   W

Laundry

Utility

30'-6''

Master bedroom
18'-5'' x 18'-4''

Bedroom
12'-0'' x 11'-7''

Bedroom
12'-0'' x 11'-7''

Mudroom
5'-0'' x 5'-0''

Skylight

Patio
14'-5'' x 12'-6''

29'-6''

46'-0''

Second floor

Storage
20'-3'' x 12'-5''

Down

Hobby room
13'-8'' x 22'-6''

30'-6''

North

32'-10''

Bedroom
12'-3'' x 18'-0''

Guest room
11'-9'' x 12'-0''

Porch
14'-5'' x 6'-6''

Exterior walls: 8'' milled log
Interior walls: 2x6 frame

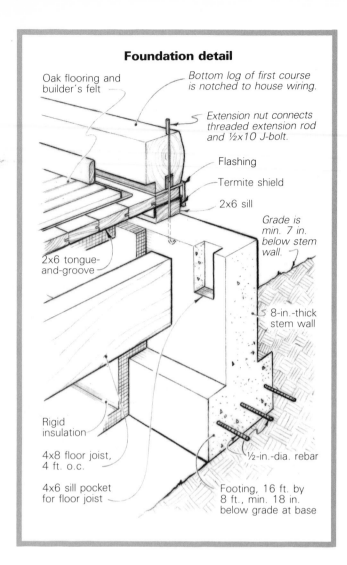

Foundation detail

Oak flooring and builder's felt

Bottom log of first course is notched to house wiring.

Extension nut connects threaded extension rod and ½x10 J-bolt.

Flashing

Termite shield

2x6 sill

Grade is min. 7 in. below stem wall.

2x6 tongue-and-groove

8-in.-thick stem wall

Rigid insulation

4x8 floor joist, 4 ft. o.c.

½-in.-dia. rebar

4x6 sill pocket for floor joist

Footing, 16 ft. by 8 ft., min. 18 in. below grade at base

The foundation

The foundation for the house is a 16-in.-wide by 8-in.-thick footing with an 8-in.-thick stem wall that extends 7 in. above grade. Three pieces of #4 rebar (½-in.-dia.) run horizontally through the footing, and *J*-bolts extend through the top of the stem wall every 3 ft. to tie the pressure-treated 2x6 mud sill to the bottom log of each wall. A vapor barrier of 6-mil polyvinyl was laid on the ground under the floor area. Covering the sill is a termite shield made of brown anodized aluminum flashing. A second layer of flashing protects the exposed outside edge of the decking, which runs out to the edge of the sill. This extends far enough to cover the termite shield, and from outside the house all that can be seen is the top flashing.

In the dining room and bedrooms, concrete piers were poured on 4-ft. centers to support the 4x8 floor joists. To create a pocket in the foundation walls for the ends of the floor joists, 4x6 plastic boxes were placed in the top of the formwork. Over the joists, Bridgwater nailed down 2x6 hem-fir tongue-and-groove decking. (She used the same decking on the 10x10 barn-beam floor joists left from the old house.) Before nailing down the oak finish-flooring, Bridgwater would put down a vapor barrier of builder's felt over the decking. Insulation in the floor is 1½-in.-thick rigid foam between the joists.

Beginning the walls

With the decking installed, Bridgwater was ready to start the log walls. The logs were delivered in 12-ft., 14-ft. and 16-ft. lengths. Although the milled logs were free of taper, Bridgwater knew there was still a little variation in thickness. Her plan was to stop at the tops of the windows to check the level of the logs and make any necessary adjustments at that point.

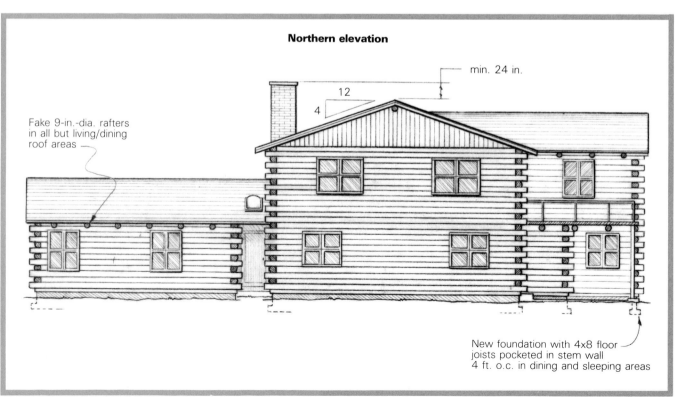

Northern elevation

min. 24 in.

12
4

Fake 9-in.-dia. rafters in all but living/dining roof areas

New foundation with 4x8 floor joists pocketed in stem wall 4 ft. o.c. in dining and sleeping areas

Before putting down the first logs, Bridgwater and her crew of two plumbed and braced all the door bucks in place. The bucks were made of 2x10 boards with a 2x2 nailed up the center of the outside of each vertical board. The logs that butt against each buck are notched to fit the 2x2 nailer, keeping air infiltration around the doors to a minimum. (The window bucks would later be handled in the same manner.) The notches on the logs were cut to fit tightly, but to allow for log movement and shrinkage the bucks were not nailed to the logs.

The logs were light enough to be carried easily by two workers, and Bridgwater and the crew stockpiled enough logs on the floor of the house to start the first course. Then they drilled the holes for the J-bolts. The ends of the logs had to be squared up before they were placed in the wall, and where logs butted together end-to-end in a wall, a slot to hold a spline of ½-in.-thick plywood had to be sawn in each end. Before laying down the logs, Bridgwater filled both slots with caulk and, with the logs butted up tightly in place, she drove the spline in with a hammer. With a circular saw, Bridgwater also cut a 2x2 channel for electrical wiring on the bottom of each log in the first course. Then she cut the holes for the outlet boxes just above the channel. The J-bolts were extended with long nuts and pieces of threaded rod. Each log was then put in place on top of the flashing that covered the outside edge of the decking.

By laying the logs right on the decking this way, Bridgwater saved the extra work of cutting sill logs and notching them for the floor joists. Also, the decking made a good work area for sliding the logs and for marking and cutting. Bridgwater was basically using the same type of floor system that's used for a stick house, and laying the first course of logs where a stick-house builder would put down the plates.

Here the window bucks are in place. All bucks have a 2x2 nailer on each side, visible at the top of the bucks, to which the logs are then notched to minimize air infiltration. The bucks are not nailed to the logs but are allowed to float to allow for wall movement.

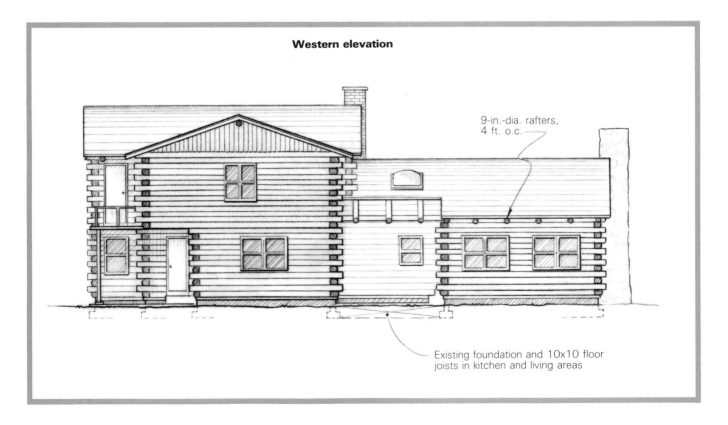

Western elevation

9-in.-dia. rafters, 4 ft. o.c.

Existing foundation and 10x10 floor joists in kitchen and living areas

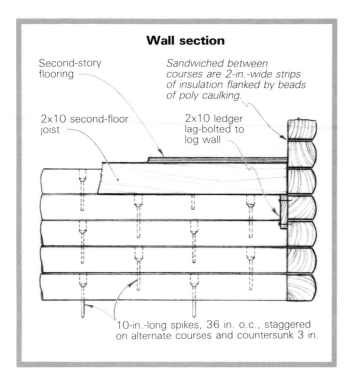

Wall section

Second-story flooring

Sandwiched between courses are 2-in.-wide strips of insulation flanked by beads of poly caulking.

2x10 second-floor joist

2x10 ledger lag-bolted to log wall

10-in.-long spikes, 36 in. o.c., staggered on alternate courses and countersunk 3 in.

The logs at the corners of the house butt up against each other with alternate logs running long. Each log is spiked into the one beneath it with two 10-in.-long spikes.

Once the bottom course of logs was in place, wall-laying began in earnest. Down the center of each log, Bridgwater ran a 2-in.-wide strip of fiberglass insulation. On each side of that she ran two heavy beads of polyurethane caulking. The next course of logs was then laid in place and spiked with 10-in.-long spikes on 36-in. centers. Bridgwater staggered the spikes from course to course so that they lined up on every other log. She drilled holes through to the bottom logs, then countersunk 3 in. to get good purchase. The spikes were driven with a 12-lb. sledgehammer and pounded home through the countersunk part of the hole with a catspaw—Bridgwater went through a lot of catspaws.

On the exterior corners, Bridgwater simply butted one course against the other, letting alternate courses extend past the edge of the building. This way she saved the time it would have taken to scribe and cut notches or dovetails. Each overlapping log is double-spiked to the one below. On the interior corners, the ends of the logs are sawn flush for a square, flat wall.

For this part of the log work, Bridgwater had a crew of four. Because the logs were in short lengths, the crew could work on one wall at a time, building it to a comfortable height before moving on to another one. After the fourth course of logs, it was time to put up the 2x10 window bucks, set so that the tops of the windows would be even with the tops of the doors. The crew was able to get all the wall logs up by hand, using step ladders as needed. Once the logs reached the window and door tops, Bridgwater leveled the whole building. She marked the level line all along the top logs, snapped a chalkline to connect the marks and power-planed the high spots—the greatest difference between the highest and lowest spots was just ⅝ in.

When the walls were 8 ft. high in the children's wing (where the second floor would be), Bridgwater lag-bolted a 2x10 ledger flat against the top logs. This is where the floor joists would sit. Once the decking was down on the floor joists, Bridgwater used a block and tackle to build up a stockpile of logs for the second-floor walls. On the south (fireplace) wall in the living room, the logs go all the way up to the top of the gable end, 15 ft. above the living-room floor. The interior log wall between the dining room and the living room is 13 ft. high and every other log in it is mortised into the logs of the south wall. The tallest section of log wall is on the north side of the house, where the wall rises 22 ft. Here, however, the gables are framed with 2x6s and half-rounds (halves of small logs, often used for siding in fake-log houses) vertically placed to break up the horizontal lines.

The roof

The living room and dining room of the house have open ceilings with exposed log rafters, beams, purlins and decking. But the rest of the house has flat sheetrock ceilings under a system of manufactured trusses set on 24-in. centers. Bridgwater felt that sheetrock ceilings would help keep the dust down, and reflect more light. In these areas, she ran some 9-in.-dia. round logs back into the attic and scabbed them onto the trusses so she could match the appearance of the log rafters in the living and dining rooms. These extend 2 ft. beyond the wall and 4 ft. back into the building, where they are bolted through to the adjoining truss and end-nailed to a piece of scrap 2x blocking between the trusses. The blocking in the eaves is notched into the fake-log rafters and covers the end of the trusses from view.

Bridgwater and her crew used two chainsaws and a homemade mill to flatten the tops of the ridgepole, purlins and log rafters. The ridgepole is a 14-in.-dia., 24-ft.-long log. In addition to flattening it with the mill, Bridgwater sawed a 3-in.-deep kerf down its center to relieve stress and to prevent twisting and checking as the log dried—the theory is that the kerf works like the expansion joint in a concrete slab. The ridgepole is mortised into the top course of logs in the south gable end. The two 12-in.-dia. purlins in the living and dining rooms are also 24 ft. long and are supported on one end by 12-in.-dia. log posts and on the other end are mortised into the south wall. These were also kerfed.

The purlins were placed by a rented crane, but just as Bridgwater and the crew were ready to set the ridgepole, the truck sunk in the mud and stuck fast. By the time they got it free, it was time to bring it back, so the next day they rigged a block and tackle on both ends of the massive ridgepole and hoisted it into place by hand.

The 9-in.-dia. rafter logs were flattened and kerfed, and a seat was cut in each one where it would rest on the top wall log. They were then set on 4-ft. centers with a 4-in-12 roof pitch and lag-bolted to the ridgepole. The rafters are covered with 2x6 pine tongue-and-groove roof decking, which is then covered by 2x4s on edge filled with two layers of 1½-in.-thick rigid insulation. There is ½ in. of dead airspace on top of the insulation with ⅝-in.-thick plywood on top. A layer of 30-lb. roofing felt and composition shingles finish off the roof.

The roof of the house is constructed from manufactured trusses. Logs 9 in. in diameter are scabbed onto the trusses to match the appearance of the log rafters in the living and dining rooms. Photo by Gary Ballou.

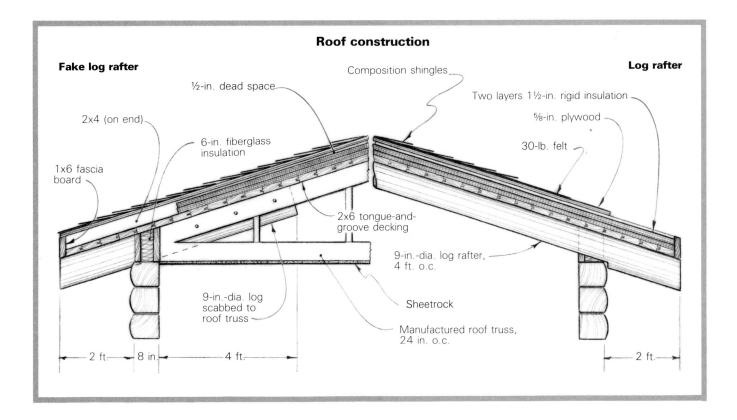

Roof construction

Fake log rafter

½-in. dead space

2x4 (on end)

6-in. fiberglass insulation

1x6 fascia board

Composition shingles

Log rafter

Two layers 1½-in. rigid insulation

⅝-in. plywood

30-lb. felt

2x6 tongue-and-groove decking

9-in.-dia. log rafter, 4 ft. o.c.

Sheetrock

9-in.-dia. log scabbed to roof truss

Manufactured roof truss, 24 in. o.c.

2 ft. 8 in. 4 ft. 2 ft.

The screened-in patio/sunroom attached to the front porch, kitchen and master bedroom has a flat roof built of metal roofing over a 2x6 framework.

The two 4x8 double-glazed bubble skylights in the living room are screwed into 2x10 fir boxes. The boxes fit between two rafters and 9-in.-dia. logs are scribed into the rafters at the top and bottom to give the skylights a nice log frame. The inside of the box is finished with pine to blend with the surrounding roof decking.

As the roof neared completion, Bridgwater had an accident that almost turned her first big project into her last. Plywood was down over the insulation, and the roofing paper was down and ready for the shingles. Bridgwater was walking on the roof over the kitchen and walked on the paper that covered the chimney cutout for the kitchen fireplace. Down she went to the kitchen floor below. She ended up in the hospital with several fractured ribs, a broken pelvis and a collapsed lung. For the next couple of months she directed construction from her bed.

Bridgwater wasn't looking forward to framing in a flat roof for the 9x9 screened patio/sunroom and attaching it to the bottoms of the pitched roofs of the porch, kitchen and master bedroom. She was worried about water collecting on the flat roof and working its way into the sunroom at the places where the roofs joined. She solved the problem and gained peace of mind by using metal roofing over a 2x6 framework. The flat roof for the sunroom was framed in first and then the metal roofing was laid down and lapped up 1 ft. over the bottom of the three adjoining roof surfaces. Then the finished roof of composition shingles was laid over the lapped metal roofing just as it would be over any kind of metal flashing. There is a 4x8 plastic skylight in the middle of the pitched roof, and the sunroom has turned out to be light, airy and free of leaks.

Settling and finishing

The settling of a log house presents some interesting challenges in dealing with windows, doors, divider walls and attached cabinets. There is no way to predict the exact amount of settling, so allowances must be made for the worst case. The ideal solution to settling would be to finish the log work, put on the roof and come back a year later to install windows and doors, divider walls and cabinets—ideal, but not very practical.

Bridgwater calculates that in the first year the walls in her home settled only about ¼ in., an indication that the system she used has worked very well. Like most log builders, she left a 2-in. opening over the doors and windows to allow for settling. But where a downstairs stud wall tied to a log wall, Bridgwater and her crew got creative. They lag-bolted the stud to the logs, but cut vertical slots around each bolt so it could slide up and down in the stud. At the bottom of the stud wall, they inserted little stacks of removable shims that could be pulled out as the wall settled. They left a ½-in. clearance at the top of the wall where it met the upstairs floor, and used threaded rod with nuts and bolts to fasten the plate of the stud wall to the floor above. Just before sheetrocking, they pulled out a couple of shims. Bridgwater was probably overzealous in her precautions against shrinkage, especially since she used bug-killed logs, but it's a lot easier to allow for settling when building a log house than it is to repair a 2-in.-gap after the house is finished. To protect the kitchen cabinets from any movement in the walls, Bridgwater bolted 2x4s with vertical slots to the wall, then nailed the cabinetry to the 2x4s.

Before sheetrocking the divider walls, Bridgwater also nailed the 4-in.-wide cove molding at the top of the wall into specially placed blocking in the upstairs floor joists or in the overhead roof trusses (depending on the location of the wall). This way the wall could move behind the trim without popping nails or warping it. The molding covers all the extra space at the top of the walls and has a nice traditional look that complements the log work and antique furnishings.

To finish the logs, Bridgwater first went over them with a wire brush chucked in an electric drill to remove dirt and expose new, white wood. Then she brushed on Olympic Clear interior finish sealer to seal the logs and bring out their color.

The windows were set into the bucks that had been installed as the logs went up. Bridgwater used leftover 2x decking to make stops for the windows. She cut off the tongues and grooves and sanded the wood, then faced it with quarter-round molding. On the outside, she made a little flashing-strip cap on the top of each window to keep moisture from getting in between the windows and the logs.

The kitchen fireplace, built on the footings from the original, is brick and its back provides a brick wall for the bathroom next to the kitchen. Bridgwater hired Laura Hinrichs, whom she had met at the local community college, to do all the cabinets. Hinrichs built all the pine cabinets for the kitchen and bathrooms in a shop Bridgwater set up on the site. Hinrichs decorated the cabinets in the children's bathroom with trees carved with a Dremel tool. She built the countertop between the kitchen and the living room out of glued-up oak with thin strips of mahogany and put it all on a base of ¾-in.-thick oak plywood. The only special problems Hinrichs had were working with the 2x4 furring system already described and doing the scribing necessary to fit the counter and cabinets around the poles that support the kitchen end of the ridgepole and the end of the purlin that runs through the living room. These posts come at both ends of the 10-ft.-long eating bar that divides the kitchen and the living room.

The master bath contains a large jacuzzi surrounded by decking with windows on two sides. Full-length wood shutters control the light that comes through the three 5-ft.-high windows, and the white tile and Varathaned log walls bounce the light around the room.

Besides the two fireplaces, heating is provided by a Lenox pulse gas furnace. This forced-air system pumps heat to all the rooms through a standard system of ducts and registers.

For those days of outdoor living in Oregon, Bridgwater has a nicely landscaped backyard and porches. One porch is upstairs off the guest bedroom and looks out onto the backyard and the woods. The other porch is at the front entry and gives the house a traditional log look. The 10x20 front porch is set off by peeled log posts on 10-ft. centers topped by a 20-ft.-long log header. The posts are mortised to the header, and the outsides of the logs are scribed to fit neatly, tightly and traditionally. The roof over the front porch is built up on the same manufactured roof trusses used in the rest of the house, and the gable end that faces the entry path is faced with vertically nailed half-round logs.

The exterior of the logs is treated with pentachlorophenol, which fights decay and insects, and CWF (Clear Wood Finish, from the Flood Company), which is a penetrating sealer. Sealers like CWF actually penetrate the surface of the wood and allow the wood to breathe. Bridgwater figures she will have to respray the logs every three years for maximum protection from moisture, fungi and insects.

Looking back on her work, Bridgwater feels pleased with the house. She had good help from her crew and worked well alongside them. Now as she finishes up the odds and ends of trimwork and touches up, she has taken on another remodeling job, but what she is really hoping for is another good-size log project.

The Bridgwater kitchen is spacious, well-lighted and nicely organized for work flow. The post at the left supports the kitchen end of the ridgepole and the end of the purlin from the living room.

The jacuzzi tub dominates the master bath, the logs of which are finished with Varathane to protect against moisture damage.

The House on Orcas Island

CHAPTER 7

The log house that Loren and Margery Gilbertson built on Orcas Island, Washington, has only 900 sq. ft. of living area on the ground floor and another 400 in the loft above, but it looks big from the outside and feels big on the inside. It sits in a clearing on the south side of the island a few minutes from the ferry landing at Orcas Village. In the design and structural sense it is simple—a big rectangle with a loft over one half, all made from hand-hewn Douglas fir logs. But visually, the house is quite interesting. The 32-ft.-long front porch with its gable roof and huge log posts and beam make the house look like a north-woods lodge. A good-size deck on the south side of the house breaks the austerity of plain log walls and provides 400 sq. ft. of outdoor living area. Inside, the floor plan is open and the ceilings are high—the ridgepole is 18 ft. above the floor. Part of the downstairs is sectioned off into a hallway, children's bedroom, laundry and bath with wood-frame walls. Counters and a Hoosier cabinet divide the kitchen, but the room is basically open to the living room and dining area. The woodstove in the living room easily heats the house, and even overheats the loft master bedroom during winter. The huge star-brace roof truss where the main beams for the loft and the supports for the purlins meet is a strong focal point. High skylights and windows in the east gable end let in a lot of light and keep the interior from being as dark and close as in other log houses. Several coats of log oil over the years have given the logs a rich, brown luster.

Gilbertson has worked in the woods as a logger most of his adult life, so it's appropriate that he chose logs as his building material (plus it's hard to imagine any other type of house sitting as harmoniously on the grassy hillside of the building site). Gilbertson obtained the logs from log-house builder Steve Kenady (see p. 50 and photos, p. 51) and contracted to have Kenady peel, saddle-notch (this process is also called round-notching) and erect them. Gilbertson did the foundation, roof, windows and doors, flooring, interior partitions and finishing.

There is an abundance of timber on Orcas Island, but island fir is often described as being a little quirky, even by the stick-house builders who use the wood from the local mills. So Kenady got the logs for the Gilbertson house in the Mt. Baker National Forest, on the mainland just east of Bellingham, Washington. These were straight, with no twist or taper. Kenady trucked the logs over on the ferry, and with a crew of three or four peeled, sized and notched the logs. When everything was right, Kenady numbered all the logs, dismantled the house and delivered it to the Gilbertson site.

The front porch on the north side of the house spans 32 ft. The beams protruding into the porch area are the squared ends of the log joists for the loft sleeping area.

The Gilbertsons' log house on Orcas Island, Washington, has 1300 sq. ft. of living area and looks like a northwoods lodge. Steve Kenady and his firm, The Real Log Cabin Company, supplied the logs and log work. Round notches join the corners, and the chinking is mortar on wire. The gable ends are framed in and covered with vertical 1x pine siding.

Above left, a Sonotube-formed concrete post is poured on each corner of the house, and pads are poured at midpoint on each wall and down the middle for log posts that will support the floor joists. Above right, the saddle-notched corners rest on the concrete posts. Photos by Margery Gilbertson.

The foundation

As Kenady was preparing the logs, Gilbertson was working on the foundation. He chose a post-on-pad foundation system. At each corner there are 12-in.-dia. reinforced-concrete cylinders cast in Sonotube forms that sit on 4-ft. by 4-ft. concrete pads. (The Sonotube is really just a giant cardboard tube set on end and filled with concrete. You see this system used frequently in concrete-bridge and freeway-overpass construction.) At the midpoint on all four sides of the house are log posts on 4-ft. by 4-ft. concrete pads. A series of 2-ft. by 2-ft. pads runs down the center of the house on 4-ft. centers; a 10-in.-dia. log post sits on each pad and supports the 6x8 floor joists, which butt together on top of the posts and span 14 ft. from the center of the outside wall.

All the pads are reinforced with ½-in.-dia. rebar. The pads on the uphill corners of the house are set right on bedrock. On the downhill corners the pads are dug down 2 ft. to 3 ft. to undisturbed soil. Gilbertson ran four pieces of ½-in.-dia. rebar in the concrete corner posts and left one piece long to attach to the first layer of logs. (He had to drill the sill logs to fit over the rebar.) He poured the corner posts at the same time that he poured the pads.

A lot of the old log houses built in the Northwest used a wood post-on-pad foundation system. The advantage is that you don't have to level the site, dig and pour a perimeter foundation and wait for it to set up—the first course of logs can be placed right away. The disadvantage is that the log posts are subject to decay and provide a pathway for insects and rot to reach the wall logs. For many builders who like the idea of a post-on-pad foundation, poured concrete cylinders are a good compromise. They're stronger than wood, and you still can avoid the digging, forming and pouring that a perimeter foundation requires.

The log work

When the logs were delivered to the site, they were lifted off the truck with an attached crane and into a pile beside the foundation. Kenady's crew had a small boom truck for moving logs around and an old crane for lifting the logs into place in the walls. Gilbertson figures that his load-and-a-quarter of wet logs weighed around 60,000 lb.

The first logs down were the two 36-ft.-long sill logs. These were set on the concrete posts on each end and a log post at midspan. Gilbertson marked the sills on 4-ft. centers, then marked a 6-in.-wide vertical pocket for each of the 6x8 floor joists and chainsawed them out. The ends of the joists fit into the pockets and are spiked in place from the outside of the log. The joists butt together on the posts that run down the middle of the house. With the 14-ft. span of the floor joists and the 4-ft. span for the floor decking, Gilbertson says he has the best dance floor in the Northwest—that's just his way of saying that the floor is a little springy for his taste and that he probably shouldn't have used such long spans.

Once the long sills were down and notched, the end sills were set in place. These 32-ft.-long logs had already been scribed and notched on the ends in the preassembly yard by the Kenady crew, using the same procedures as described in the introduction to this section. The overlapping corners were held in place by the notches and the rebar protruding from the concrete posts on the corners. With the sill work complete and the floor joists nailed in, Gilbertson laid down the 1⅛-in.-thick tongue-and-groove plywood subfloor and the rest of the log work commenced.

Kenady did not scribe the lengths of the logs for a chink-less fit, but instead devised his own system to close the gaps between logs. A ⅜-in.-wide by 2-in.-deep slot cut on the top and bottom of each log would house a ⅜-in.-thick by

Here, the sill logs are placed and the 6x8 floor joists have been notched in. The rebar in the corner posts protrudes through the logs. A log post supports the uphill (east) sill log at midspan. Photo by Margery Gilbertson.

6-in.-wide plywood spline; chinking would later be applied on both sides of the spline—mortar on the outside and wood or mortar on the inside. As each wall log was set in place, the plywood spline was jammed in place and then trimmed along the top edge to fit the slot in the bottom of the next log.

Once the walls were over 6 ft. high, the house looked like a big log box with neatly interlocking corners. On the ninth log, the crew cut square notches into the top on 5-ft. centers to accept the floor joists for the loft. The ends of the joists, which were squared, protrude about 8 in. beyond the outside of the wall logs. A tenth log was then laid on the notched one, and the last log was put in place. Kenady used the top log on the long walls as a purlin and ran it out 4 ft. on each end of the house for the overhang. By moving the log toward either the inside of the house or the outside, he was able to do a little fine-tuning on the pitch of the roof.

At the same time that they notched for the loft floor joists, the crew cut a rough hole for the front door with a chainsaw. Up to this point the only way to get in and out of the box was with a ladder on the outside and another on the inside. The other doors and windows would be cut once the roof was on.

Because door and window openings aren't cut until the roof is on, during construction the house looks like a big log box (above right). The floor joists for the loft are squared on the ends and notched into the ninth log (right). The ends, which stick out, will be protected by the long roof overhang. Photos by Margery Gilbertson.

With all the wall logs up, the crew was ready to raise the ridgepole and the last two purlins. (Remember that the top log on each long wall also serves as a purlin.) To support these beams, six posts were scribed and toenailed to the east and west walls. The posts for the 16-in.-dia., 40-ft.-long ridgepole were raised, plumbed and braced first, then the ridgepole was put on. Pieces of decking attached to the ridgepole were run down to the purlins already in place for bracing. The posts for the last purlins were then set, and the purlins were raised into place—the strips of decking helped make their placement exact. Neither the ridgepole nor the purlins were kerfed (which some log builders believe assists more even drying) because Gilbertson feared that moisture might collect in the kerfs.

As soon as the purlins and ridgepole were secure, Gilbertson began to nail down the 2x6 tongue-and-groove decking in the loft. On the kitchen side, he ran the decking out another 2 ft. past the edge of the last loft joist. This gives 17 ft. of floor space from the end wall to the edge of the loft on that side. (On the other side, the stairs take out 4 ft. of that floor space.)

Under the edge of the loft and square in the middle of the last 32-ft.-long loft joist, Gilbertson set a vertical log post. At the same point, but on top of the loft decking, he set another log post to hold the ridgepole at its midpoint. From the same spot on the decking he ran two logs out to support the purlins, creating a giant truss. It was places like this that Kenady's crane came in handy. Underneath the loft, at the same angle, a log scribed into the top of the first-floor post points toward the front entrance. This latter log, which doubles as the stringer for the stairs up to the loft, was tricky to scribe and match. It had to be scribed on the horizontal to butt up against the loft joist and on the vertical to match the post. There was also the angle at which the stringer came

into the junction to cope with, as well as the bevel cut on the log's floor end. When it came time to build the stairs, Gilbertson first mortised each tread into the stringer and then fastened the other end to the 2x6 divider wall that runs under the loft joist. The rough cedar paneling that covers the divider wall is cut and fit around each planed and polished tread for an effective contrast of color and texture.

Gilbertson's original plans called for the loft joists to be on 4-ft. centers, but as the construction proceeded, he decided to make the loft 15 ft. long instead of 12 ft. This changed everything to 5-ft. centers, which was fine except that he forgot that he was getting a mismatch under the floor, where the first-floor floor joists were supported on 4-ft. centers. The day of reckoning came a day or two after the massive star bracing was put together. Gilbertson and his father were sitting under the house in the shade having a smoke, when Gilbertson looked up to see what looked like a huge punch coming through the 1⅛-in.-thick plywood. It was the bottom of the post support for the ridgepole and it was now resting on a piece of unsupported plywood floor between the 6x8 floor joists. Luckily, they were able to scab in another post at that point to relieve the stress before any real damage occurred, then they went on to correct the problem.

Gilbertson now began to nail down the roof decking. Some people use rafters *and* purlins in log construction, but Gilbertson saw no need for this. He simply ran 20-ft.-long pieces of 2x6 tongue-and-groove decking from the ridgepole to the walls with an overhang of 3 ft. 8 in. on the sides. With the purlins running out 4 ft. on the gable ends, this gave Gilbertson a generous overhang all around the house and good protection for the log walls against any kind of weather. The decking spans 8 ft. from ridgepole to purlin and another 8 ft. from purlin to wall top.

The end posts for the ridgepole and the purlins are mounted on the top log on the west and east ends of the building. On the facing page, the crane is setting the 40-ft.-long ridgepole, which will be spiked to the post with a rebar pin. Above, the ridgepole and purlins are braced in place with pieces of roof decking. All posts have been scribed on both ends for a tight fit. Photos by Margery Gilbertson.

The central star truss for the roof system is a dramatic focal point and divides the back sleeping/bath/laundry area from the front living/dining/kitchen area. The log on the lower level to the left is the stringer for the loft stairs. The scribed cut on the stair log stringer is complicated by the meeting of three logs and the angle of the stringer.

The front-porch beams are set on concrete pads and log posts just as the rest of the house is. A ledger bolted to the bottom log of the house wall receives nailing for the decking. Photo by Margery Gilbertson.

The ridgepole and purlins of the porch roof are braced and ready for decking. The house roof is partially decked, and the beveled ends of the porch ridgepole and purlins have been spiked into the decking. Photo by Margery Gilbertson.

Gilbertson started the front porch as the roof decking was going on. He set down six more 2-ft. by 2-ft. concrete pads with log posts to support the beams, which were laid parallel to the front of the house on 5-ft. centers. He ran the 2x6 porch decking at right angles to the house and nailed it onto a 2x10 ledger strip lag-bolted to the bottom log of the house wall. With the decking down, he erected the two 10-ft.-long corner posts for the porch roof. Where the 20-ft.-long beam is supported by the posts, Gilbertson power-planed two flat spots on the beam and then spiked it onto the posts with rebar spikes. The gable roof covers the center 20 ft. of the porch and is supported by a log-post framework. The house end of the porch ridgepole is cut to match the slope of the main roof and is then spiked right to the roof decking. The two purlins at the eaves are handled in the same manner.

On top of the house roof decking, Gilbertson laid down 1 in. of styrofoam for insulation with a layer of 30-lb. felt over it. For his finish roofing, he used cedar shakes, but says he would try metal roofing if he were ever to re-roof. Gilbertson hasn't had the trouble that some Northwest builders have had with carpenter ants; apparently, the ants think styrofoam is the greatest thing since cotton candy, and some people who've used it claim they can hear the ants chomping on it at night. I don't know if they eat it or just enjoy being around it, but I've heard from several builders that when you use styrofoam, it's like inviting the carpenter ants to move in.

With the roof on, the crew chainsawed out the rest of the door and window openings. For the doorways and the two large windows in the living and dining rooms, they used a 2x6 guide and a chainsaw mill to get the cuts square. They also cut the two windows in the children's room and a small window over the kitchen sink. Upstairs on the east wall are tall windows on each side of the bed; these were framed in with the stud framing of the gable ends. Both ends were framed, covered with plywood and insulated. On the outside, Gilbertson used vertical 1x pine siding to break up the lines of the horizontal logs. Twin skylights were also framed into the roof above the loft bedroom.

Once the openings were cut, Kenady kerfed both sides of the opening 2¼ in. deep with a circular saw and then pounded in a length of 2-in.-wide angle iron the length of the opening on each side. He set the angle iron ¼ in. into the logs so that as he spiked through on each log end with his 80d spikes, the angle iron would be flush with the log ends and invisible from the sides. The window and door trim fit tightly against the logs.

The bottom of the window openings is chainsawed at an angle for drainage. Photo by Margery Gilbertson.

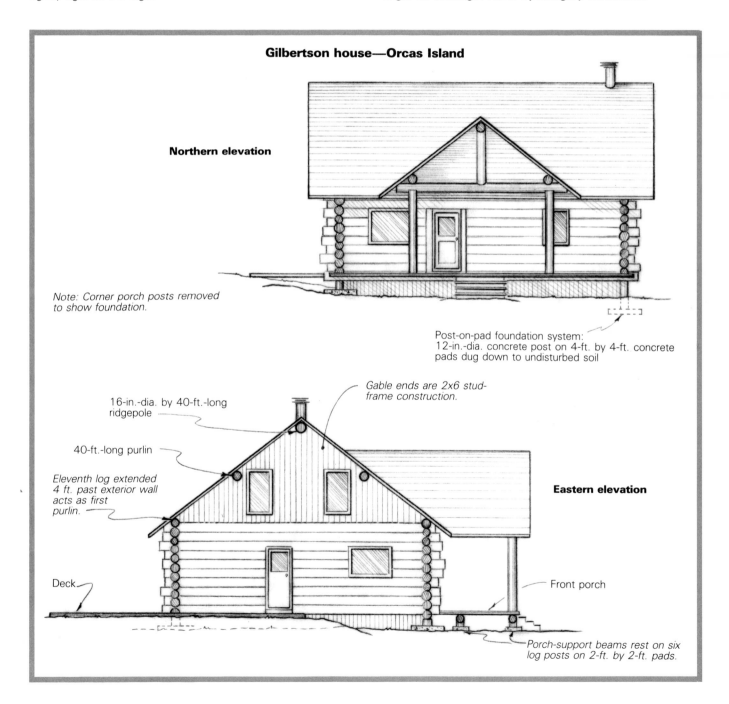

Gilbertson house—Orcas Island

Northern elevation

Note: Corner porch posts removed to show foundation.

Post-on-pad foundation system:
12-in.-dia. concrete post on 4-ft. by 4-ft. concrete pads dug down to undisturbed soil

Gable ends are 2x6 stud-frame construction.

16-in.-dia. by 40-ft.-long ridgepole

40-ft.-long purlin

Eleventh log extended 4 ft. past exterior wall acts as first purlin.

Eastern elevation

Deck

Front porch

Porch-support beams rest on six log posts on 2-ft. by 2-ft. pads.

Finishing up

With the house closed in, Gilbertson laid down the oak flooring and installed the kitchen cabinets. He also built interior stud walls to create a bedroom for the children in the northeast corner and a bathroom in the southeast corner. The back door opens between these two rooms and the laundry area is set in against the bath side. This makes a hallway into the kitchen and living area.

Plumbing for the house is hidden in the interior stud walls, and vents and the kitchen-stove chimney are run through the loft storage area. Electrical wiring is also run in the stud walls except for the outlet boxes routed into the bottom logs around the floor. The wiring for these is drilled through from the bottom of the log and is not visible in the rooms.

The family moved into the house in 1980, but Gilbertson didn't get around to chinking the house until 1984 when some friends were visiting and offered to help. He used a standard prepackaged mortar mix, but a problem with mortar chinking is that it doesn't shrink and swell along with the

logs, making it liable to pop out. So Gilbertson wired the openings between the logs with a single strand of wire in a zigzag pattern with bent-over 8d galvanized nails every 3 in. to 4 in. to hold it in place. (The nails are bent in toward the logs so they don't show when the mortar is applied.) The mortar joint is about 2 in. to 2½ in. high and extends back to the plywood spline. In this type of log work, the mix must be stiff yet wet enough to be worked between the logs and around the wire. It can be applied with a narrow trowel or even a 2-in.-wide putty knife. As the mortar dries to the sandy, crumbly stage, the edges of the joints and the adjoining logs must be cleaned off with wire brushes. By the time the Gilbertsons chinked their walls, they had already applied several coats of log oil to the logs, so the excess mortar was easily cleaned off. Today the mortar shows signs of cracking horizontally where it meets the log, but the wire is holding it in place. This is where the plywood spline is important, because even as the mortar breaks its seal, the spline blocks air infiltration.

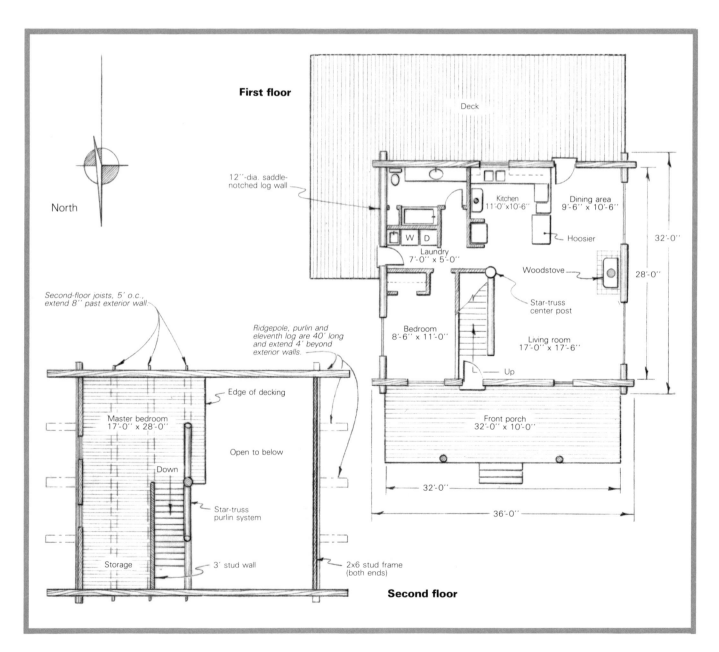

First floor

North

12''-dia. saddle-notched log wall

Deck

Kitchen
11'-0''x10'-6''

Dining area
9'-6'' x 10'-6''

Hoosier

W   D

Laundry
7'-0'' x 5'-0''

Woodstove

32'-0''

28'-0''

Second-floor joists, 5' o.c., extend 8'' past exterior wall.

Star-truss center post

Bedroom
8'-6'' x 11'-0''

Ridgepole, purlin and eleventh log are 40' long and extend 4' beyond exterior walls.

Living room
17'-0'' x 17'-6''

Edge of decking

Up

Master bedroom
17'-0'' x 28'-0''

Open to below

Front porch
32'-0'' x 10'-0''

Down

Star-truss purlin system

32'-0''

Storage

3' stud wall

2x6 stud frame
(both ends)

36'-0''

Second floor

The log-oil mixture Gilbertson initially brushed on the outside of the house consisted of a ratio of 5 gal. of log oil to ½ gal. of mineral spirits and 1 qt. of pentachlorophenol. (Recently he gave the logs another application and found that the south wall really soaked up a lot.) He used nearly 30 gal. of the mix. On the inside he used one coat of log oil (without the pentachlorophenol and mineral spirits) heavily thinned with diesel fuel to avoid the tacky surface that log oil sometimes leaves.

Now that the Gilbertsons have been living in their house for over six years, how is it working out? They like it fine, and have found that they can live with the special housekeeping demands of logs and a woodburning stove. Pointing to the massive upright log in the top of the star bracing, Gilbertson showed me where the weight of the roof and the huge ridgepole appear to be unwinding the grain in the post. Nothing is really going anyplace, but it is an instructive lesson in the weights and stresses involved in a house with this much mass. Gilbertson also showed me a spot in the laminated oak counter that divides the kitchen stove from the dining area where the lamination has popped open about ½ in. He feels that this is because the downhill corner of the house has settled more than any other part.

Gilbertson finds that the logs in the house pick up moisture in the summer when the woodstove is mostly off, and dry out in the winter when the wood heater is cooking night and day. Even so, the house stays comfortable year round.

The floor is insulated with 1-in.-thick styrofoam, and the skirting all around the crawl space is thoroughly caulked and plugged and also insulated. There is no chinking on the inside of the house, but Gilbertson plans to stuff the corners, which open as the logs go through their summer/winter cycles, with oakum.

On my most recent visit to the Gilbertsons in the summer of 1985, I was shown their newly installed front door. A local woodworker had carved the duckhead handle and fashioned the wooden latch. Sue Joyner, a neighbor on the island (see p. 49), built the beveled-glass window. After six years, the Gilbertsons' house, like most owner-builder projects, is still being finished.

A visit from some of the Gilbertsons' friends ended up as a chinking party. The concrete-mortar joints are wired with single-strand wire on nails in a zigzag pattern. The logs have a heavy coat of oil, so mortar can't soak into their faces. Cleanup was with wire brushes and scrapers.

The wooden front door has custom glass work and a duckhead handle.

The House at Grizzly Flat

CHAPTER 8

Of the three log houses discussed in detail in this section, the house described here is the most difficult style to build, involving as it does the sophisticated notching and scribing techniques of Scandinavian, or chinkless, log building. It was built by John Lee, director of the Canavi Log Home Company in Somerset, California, for Sue and Dennis Densmore, also of Somerset. The house is not the fanciest or the most unusual one that Lee has done, but it is the most popular floor plan that he builds.

Up until this project, Lee had done only the log work on his houses—the client was expected to choose his or her own contractor for the foundation, floors, roof, interior divider walls, plumbing, electrical wiring, fixtures, cabinetry, doors, windows and other finish carpentry. In this case, the Densmores had been pretty much obligated to hire the contractor recommended by the bank that gave them their loan (the bank wanted someone with a track record to sign the papers so they could be sure the house would be finished and would have the value that had been lent on it). But the contractor put the roof decking on wrong, hot-mopped the bathroom wrong and put in the windows wrong. The Densmores halted the job and asked Lee if he would correct the mistakes and finish the house for them. He agreed. He also became convinced that it was important to have a general contractor's license and he received his final papers in time to finish the Densmores' house properly. (Lee is now training his own finish carpenters, and with his contractor's license he can hire tradesmen who will finish the houses to his specifications.)

Sue and Dennis Densmore's 1800-sq.-ft. log house was built by hand-hewn log builder John Lee in the Scandinavian, or chinkless, style. The upstairs dormers give the bedrooms more space and provide lots of natural light. The concrete foundation is covered with stone for a more natural look.

Lee, a transplanted Canadian, constructs his buildings from hand-peeled, hand-scribed and notched logs, preassembles them in the company's yard, then delivers and erects the home on the client's site. He builds 10 or 12 houses a year this way, many of them in El Dorado County, California. He has his share of stories to tell, as does anyone who builds anything the least bit out of the ordinary. His experience with the building department reminds me of a story told by David Easton, who built the rammed-earth house covered in Chapter 10. Easton went to the building department in the San Francisco Bay Area with a set of plans in hand. The official took one look at them and said incredulously, "Earth? Cement? I'll tell you one thing, you'll never build that here." Easton's engineer later called the official to say that he had a client who wanted to build a reinforced-concrete post-and-lintel structure with stabilized soil infill. And the official said, "No problem." I was surprised to hear Lee repeat a similar story about another California building department, because while you would expect someone to be leery of rammed earth, it's almost unpatriotic to question logs. Lee's engineer got around the problem by referring to the proposed log house as a "solid timber-frame building." (The lesson here is obvious: When dealing with building departments, watch your language.) It's interesting that once Easton's home-county building officials became familiar with his work, they became one of his best promoters. Lee has found that the same thing has happened with his local building authorities.

The Densmores chose log for its durability and strength, not for its rustic look. The house is carpeted and tiled, with the type of fixtures and furnishings most mainstream American families would be comfortable with. Sue Densmore cringes when John Lee threatens to bring in a moose head for over their mantle, or when he jokes about putting crossed snowshoes on the front door. It's not that kind of house.

The Densmores paid about $80,000 for their 1800 sq. ft. of log-enclosed comfort—about the same they would have paid for a tract home. (The logs themselves, and their as-

The dining room is comfortable and warm, but by no means rustic. Two 6-ft. windows provide plenty of light.

sembly upon the site, cost about $24,000; the mistakes made by the first contractor added up to $10,000 of the total.) Despite the fact that this design is the smallest that Lee builds, the rooms in the Densmore house are large, and the exposed purlins and ridgepole of the living room ceiling make for a dramatic entry area. The dining room and kitchen are each brightened by two 6-ft.-wide windows. The stairway is especially nice, built from polished stringers, railings and pickets made of trimmings from the log sills and purlins. The three-step landing ties the stairway and exposed upstairs gallery into the living room. The huge log truss above the stairs and gallery echoes the massive wall logs and gives this small house a big feeling. Outside a partially covered porch wraps around the north side of the house to provide extra living and entertaining area. Several coats of wood coating have preserved the fresh, yellow, just-peeled look of the white fir logs, and the stone-covered concrete foundation walls anchor the house visually to the site.

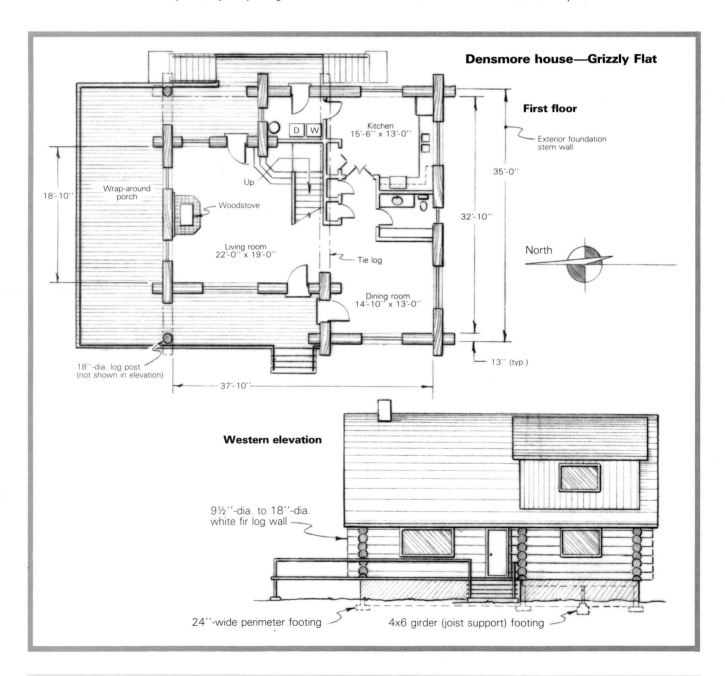

Densmore house—Grizzly Flat

First floor

Exterior foundation stem wall

Kitchen
15'-6" x 13'-0"

35'-0"

32'-10"

North

Wrap-around
porch

18'-10"

Up

Woodstove

Living room
22'-0" x 19'-0"

Tie log

Dining room
14'-10" x 13'-0"

13" (typ.)

18"-dia. log post
(not shown in elevation)

37'-10"

Western elevation

9½"-dia. to 18"-dia.
white fir log wall

24"-wide perimeter footing

4x6 girder (joist support) footing

The view from the downhill side of the Densmores' house shows the covered porches on both sides of the living room. The gable ends are framed in and covered with 1x6 siding.

The living room is open to the purlins and roof decking. The log truss and stairway create a visual centerpiece at the middle of the house. The broad landing on the left creates a dramatic base for the stairway.

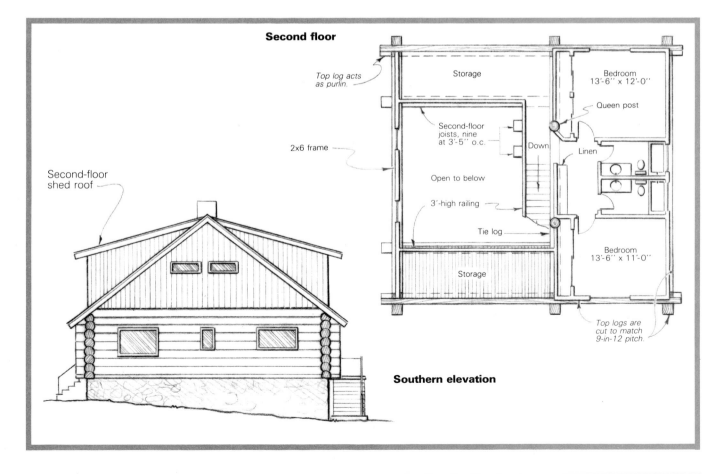

Second floor

Top log acts as purlin.

Storage

Bedroom
13'-6'' x 12'-0''

Queen post

Second-floor
shed roof

2x6 frame

Second-floor joists, nine at 3'-5'' o.c.

Down

Linen

Open to below

3'-high railing

Tie log

Bedroom
13'-6'' x 11'-0''

Storage

Top logs are cut to match 9-in-12 pitch.

Southern elevation

Getting started

The Densmores' house began as trees standing in the nearby forests of the Sierra Nevada mountains. As he does for all his houses, Lee chose the trees himself, then had them cut and delivered to the assembly site (about 10 miles from the Densmores' building site). Lee favors white fir logs, though two species of pine, Douglas fir and cedar are available in the area. Local loggers derisively call white fir trees "piss fir," treating them as good-for-nothing junk, but Lee explains that the reason he likes white fir so much is that it is all sapwood, unlike its Douglas fir cousin, which has a core of heartwood. This means that a white fir log will shrink slowly and uniformly, while the 2-in. to 3-in. band of sapwood on the outside of a Douglas fir log will shrink much more rapidly than the heartwood, resulting in checks and splits. Many log-building systems have been designed to get around the fact that sapwood is less stable than heartwood. Square logs with dovetail joints are basically all heartwood and have a long life even when untreated because heartwood is more decay-resistant than sapwood—an important consideration when preservatives and sealers were virtually unknown and unavailable. With a good roof overhang and sealer, Lee doesn't worry about decay, but instead stresses the excellent insulative qualities of white fir logs—the large cells of this wood hold more air than the small, tightly compacted cells of denser wood.

The crew peels the logs by hand with 18-in. drawknives and peeling spuds (a tool that looks like a big chisel; see photo, p. 53). The drawknives are handmade, as are the trimming spuds, which are made from automobile leaf springs. Once the logs are peeled, they are wedged. In the introduction to this log section, I discussed the way some log builders saw a kerf down each log to prevent checking and promote shrinkage along prearranged lines. Lee has an interesting variation on this procedure. He drives small oak wedges (3 in. long by 2 in. wide by ⅝ in. thick) in a straight line along the top of each log on 3-ft. centers. This causes a

Lee uses hardwood wedges to promote checking along controlled paths. The ends of these white fir logs show the checking pattern he looks for as a verification of proper log placement. The major check at the bottom of each end will be matched by one at the top as the logs dry and cure.

check that can be seen on the top of each log end perpendicular to the ground. Because Lee places the logs crown-up in the walls, which eliminates any tendency for the logs to roll in one direction or the other, a check is produced on the bottom of each log end that is in line with the top check. This pattern of double-checking gives Lee a visual verification that all the logs are set correctly in a wall. The logs in the photo on the facing page have the check at the bottom and just a hairline at the top, but Lee assured me that the latter would open as the log continued to dry and shrink.

At the building site, a foundation system is poured and a deck built up on concrete stem walls with conventional framing materials—the deck is the same as would be used for a typical frame house. To support the approximately 85,000 lb. of logs, footings are 24 in. wide with two pieces of rebar along the bottom. Pads are poured to support 4x6 wood posts, which support 4x6 floor girders. The 8-in.-thick stem wall, which is capped with a pressure-treated 2x6 mud sill, has ¾-in. anchor bolts set in at 6-ft. intervals. Joists are made of Douglas fir 2x8s, and the subfloor is ⅝-in.-thick plywood.

Back at the yard, the first sill log is pulled from the pile and sawn flat to produce virtually half a log by the time it is finished. The flat bottom will lie right on the subfloor. During preassembly at the yard, however, the half-logs are laid up on a "foundation" of leveled log sections. To cut the half-logs, a line is snapped down both sides of the length of the log and then foreman Neal Nygard chainsaws the cut free-hand. In the Densmore house, half-logs are used on the two end walls and the short walls on the front and back porches. The connecting sill logs also must be notched and sawn flat, but because they ride on the half-logs, less wood is sliced off. When the house is delivered to the site, the sill logs are laid out along the edges of the subfloor and holes are drilled on 6-ft. centers for the ¾-in. bolts that tie them to the foundation. Fiberglass insulation is laid down on top of the plywood subfloor under this course of half-logs.

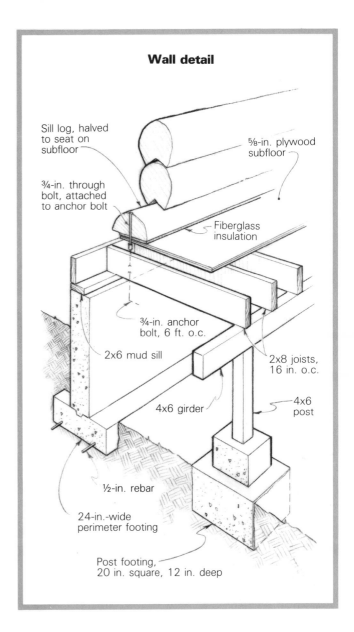

Wall detail

Sill log, halved to seat on subfloor

¾-in. through bolt, attached to anchor bolt

⅝-in. plywood subfloor

Fiberglass insulation

¾-in. anchor bolt, 6 ft. o.c.

2x6 mud sill

2x8 joists, 16 in. o.c.

4x6 girder

4x6 post

½-in. rebar

24-in.-wide perimeter footing

Post footing, 20 in. square, 12 in. deep

In the bottom photo on the facing page, the foundation and deck of the Densmore house are in place. Insulation is laid out along the perimeter in preparation for the sill logs. At left, the half-log sill is down. The notched sill is being put in place with the help of the crane. Photos by Carol Lee.

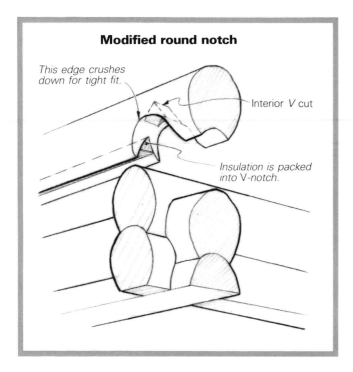

Modified round notch

This edge crushes down for tight fit.

Interior V cut

Insulation is packed into V-notch.

Three tools are used to scribe the logs. At top, Neal Nygard holds the micro-scribe that he helped develop at John Lee's firm, the Canavi Log Home Company. He adjusts the scribe to the size of the gap between the logs by changing the length of the graphite lumber crayon in the holder at the right end of the tool. Above are two compass scribes designed with marking ends reversed from one another.

Building the walls

Lee builds in the chinkless style described in the introduction to this section. He uses modified round corner notches, cut so that the thin edge of the notch will be crushed into the bottom log for the tightest possible fit. The *V* of the corner notch runs perpendicular to the *V* that runs the length of each log. The *V*-notch along the log stops short of the ends of the log and is packed with fiberglass insulation.

Many of the logs for the Densmores' house are 16 in. to 18 in. in diameter. Those on the back wall are close to 40 ft. long. The walls are eight logs high and the house has eight corners, which means a lot of scribing and cutting of corner notches. The huge log truss is made up of 18-in.-dia. rafters, queen posts, and a collar tie scribed and notched together. Seven upper-floor log joists are joined to the tie logs with double-scribed square notches, one of the trickiest notches to cut in round-log work.

Lee uses three separate scribing tools for the log work: a micro-scribe and two regular scribes that are designed with marking ends reversed from each other. The micro-scribe is an ingenious tool that Canadian log builders might recognize—Neal Nygard designed it after a prototype pipe scribe that Lee brought to California from his Canadian log-building days. The original was made from 12-in.-long, ½-in.-dia. copper tubing with a *T* soldered to the business end. One leg of the *T* was sawn off and the other leg had a little set-screw in it to hold pieces of graphite lumber crayon. By varying the lengths of the piece of crayon, Lee could scribe anything down to ¼ in. The micro-scribe is primarily used for scribing the *V*-notch along the bottom of the length of each log, and Nygard points out that its real advantage is that if you make a mistake in scribing and have to do it again, the micro-scribe will fit in the tight space between the two logs, whereas the scribe that looks like a compass won't. Nygard used a longer (1½-in. or 2-in.) crayon to make the original scribe. He has polished the idea of the pipe scribe into a neat-looking, aluminum-handled instrument complete with level. The heel of the *L*-shaped tool rides along the top of the bottom log, tracing the contours onto the bottom of the top log. The reason for two separate compass-style scribes is that the level on each one is mounted in such a way that it is unreadable if the scribe is turned over. Lee needs the reversing scribes in two places in the log-building process: once for his double-scribed corner notch and again for the complicated joist notches described on p. 90. On the corners Lee scribes and cuts both bottom and top logs for an especially tight fit. (Another trick Lee uses is to cut the corner notches about ⅛ in. overtight and then to drive them home with a heavy hammer.)

When Lee scribes a log, he first sets it in place in the wall, measures the distance between the logs and sets the scribe for a first cut on the corners. This cut is rough—just enough is taken out for a rough fit that still leaves some airspace between the logs. If the log below is crowned and the log above is straight, it's important not to make these first cuts too deep, or the top log will roll around on the crown and will have to be blocked with some 2x4 scraps so it doesn't roll off the wall. With a crane it's fairly easy to take a log back down for cutting, but sometimes if Lee is using a full-length log and it is not monstrous, he rolls the log over onto the end walls or even onto the steel-pipe scaffold he built as a log rest to do the work. With big logs you've got to be careful not to lose one off the wall, as it could do some real damage.

Lee uses short logs for walls with window and door openings rather than cutting in the openings later as discussed in Chapter 7. The ends of the logs are mortised the width of a 2x4, and once the logs are in place, a 2x4 spline is nailed into the mortise to hold them in alignment. Corners built from short logs are moved as a unit to the building site to save time in disassembly and reassembly. Held together with tacked-on scrapwood, the logs are hoisted into place by a crane. If window and door openings are done squarely, window and door trim can be kept to a minimum size; sloppy work at this point would call for 3-in. to 4-in. trim nailed over the mistakes. Lee takes special pride in tucking doors and windows tightly and exactly into their rough bucks, caulking thoroughly and trimming them out with 2-in. stock that goes right to the edge of the logs. The Densmores' windows had to be completely redone because the first contractor fitted them so badly that 4-in. and 5-in. trim would have been needed to run out over the adjoining logs and cover up the gaps.

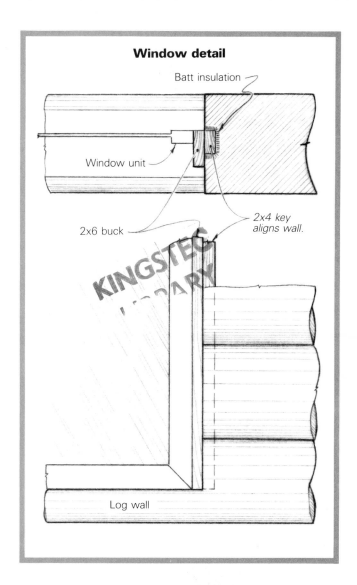

Window detail

Batt insulation

Window unit

2x6 buck

2x4 key aligns wall.

Log wall

The walls go up. The short logs at the left mark the opening to what will be the dining room. The slot in each log end will be filled with a vertical 2x4 to hold the logs in place. Photo by Carol Lee.

Preassembled corners made of short logs are moved as a unit to the building site. Scrapwood is nailed on to hold the logs together as the crane lifts them into place. This saves the time it would take to disassemble then reassemble. Photo by Carol Lee.

A 3-in. gap is left at the tops of all door and window openings to allow for settling. The top trim will be nailed to a nailer on the underside of the log header. Photo by Carol Lee.

A similar gap at the top of the interior sheetrocked walls allows for settling as the logs shrink. Four-inch trim board is nailed to the 2x4 nailer after the walls have been painted. Photo by Carol Lee.

This is the square notch used for attaching floor joists. The top wall log has a squared hump over which the joist is laid. Each log is scribed to the contour of the other and the squared portions lock together for superior strength.

When building the door and window openings, Lee first sets the 2x6 bucks in place. The top 2x6 is set 3 in. down from the bottom of the log at the top of the opening. This allows for settling so that the windows won't break and the doors will continue to open. Some builders call for a ¾-in. gap for each 1 ft. of log wall—this would require a 6-in. gap for an 8-ft.-high wall. With the white fir Lee uses, and the fact that his logs are at least partly dry by the time the house is finally assembled, Lee finds the 3-in. gap adequate. The trim that covers this gap is top-nailed to a nailer attached to the bottom of the header log. It is not nailed to the door or window jamb, so when the logs settle, the trim can move freely. Inside the house, the sheetrocked divider walls also have a 3-in. gap between them and the ceiling. These gaps are covered by 4-in. trim that has also been top-nailed only.

The fanciest notching in the house is done on the ends of the floor joists on the second floor. It's important that all the joist tops be level, so Lee takes a reading with a transit from the top of the wall for each end of each 18-in.-dia. joist to tell him how deep the notches have to be. Once both ends of each log are marked, the notches are marked. Because this is another case where the bottom log is also scribed and cut, the two reverse-end compass scribes are used here. Then a square notch is cut within the scribed areas of the log ends to tie the opposite walls tightly together. The

bottom log is left with a block-shaped hunk in the middle of the notch, and the top log is carved out squarely with saw and chisels to slip over it and tightly lock. Nygard says that this is the most complicated notch they do on a regular basis.

Before moving the Densmore house to its site, Lee assembled the roof system and cut the purlins. Each of the top wall logs that runs parallel to the eaves acts as a purlin and is chainsawn at a slant to match the 9-in-12 pitch of the roof. The ends of the top logs at the gable ends are given the same cut. The tops of the ten 20-in.-dia. purlins are also chainsawn to match the roof pitch. The 22-in.-dia. ridgepole is spliced where it crosses the truss in the middle of the house and steel is strapped on both sides. The truss itself is made up of 18-in.-dia. logs mortised and bolted together. The rafter ends are mortised into the tie log that spans the width of the house. This truss creates an interesting focal point above the stairs and at the edge of the second floor with its two queen posts (mortised on both ends) and collar tie. To keep the framing easier on the living-room gable end, Lee decided not to use another log truss, but instead planned a 2x6-frame system to support the ridgepole and purlins. The upper level of the other end of the house, containing bedrooms and bath, is all conventionally framed. Shed dormers on each side give the two upstairs bedrooms full 12-ft.-high ceilings.

The top wall logs that run parallel to the eaves act as purlins. The top of each log is cut to the 9-in-12 roof pitch, as are the ends of the top logs that run parallel to the gables. The purlins over the living room area are bolted to the log truss on the right and supported by a 2x6 framework on the left. Photo by Carol Lee.

At right, the log truss supports one end of the purlins that span the open-ceilinged living room. It is made up of 18-in.-dia. logs and is mortised and bolted together. Photo by Carol Lee.

Here the purlins are in place in their 2x6 frame system and have been leveled with blocking—only the ridgepole remains to be set. Photo by Carol Lee.

Finishing up

When the house is reassembled on the site, fiberglass insulation is stuffed into the *V*-notch that runs the length of each log. (During preassembly, the logs are simply cut and stacked right to the top of the wall.) Screw jacks keep the effects of settling at bay. They are positioned on top of the log posts at the front corners of the building, on the corners of the two porches. As the walls settle, they can be screwed down to keep the sill logs, which span the full width of the house, from bowing up at the ends.

The roof is finished with 2x10 rafters on 24-in. centers over the purlins and 4x12 rafters on 16-in. centers over the bedroom dormers. Insulation is R-30 throughout the roof and R-19 in the framed walls and gable ends. The inside finish is 1x6 V-rustic decking and the outside is cedar shingles over ½-in. plywood. The decking should have been laid down parallel to the rafters, but the Densmores' contractor had not done a log house before, so he cut and fit all the decking around the purlins and parallel to them. The Densmores left it that way.

Over the years, Lee has experimented with lots of sealers and coatings for the outside of the logs, and he now uses Sikkens Cetol HLS. Designed as a three-coat system, Cetol contains a fungicide to prevent the growth of mildew and decay fungi. Cetol will probably last longer than most other coatings, but according to the manufacturer will still have to be renewed every four or five years. (Lee figures ten years for a good application on dry logs.) The Densmore house stood for a winter in the yard before Lee could get it out to the building site, and the unprotected logs started to turn black from exposure to the weather, so before dismantling it and loading it on the truck he rubbed the logs down with a 5% bleach solution to bring back their nice white-yellow color. On the inside of the house, the Densmores used a golden oak stain with a sprayed-on lacquer finish coat.

Lee doesn't seal the ends of the logs, because he feels that it's important for them to breathe. He thinks of the logs as a natural humidifier, absorbing and releasing moisture from the house and the outdoors and equalizing it through their ends. There is some evidence that a log will even increase its R-value during extremely cold weather by increasing its thermal resistance.

The Densmores have chosen some nice contrasting textures and colors to offset the logs. Inside, the bathrooms and kitchen are tiled. Sheetrock walls are wallpapered with traditional-looking designs. Outside, they have covered the skirting around the foundation walls with a rock facade. This ties the building visually to the ground and keeps it from looking like it is floating. The rock also breaks up the strong horizontal lines of the log work.

Screw jacks can be adjusted as the house settles to keep the log posts from putting undue stress on the cross beams.

The kitchen of the Densmore house is tiled and wallpapered with a traditional print.

Everyone who gets one of Lee's houses also gets the famous John Lee picnic table. Built from the same large logs as the house is, it's strong enough to support a good-size Mack truck.

What does a log-home builder think about at night? Taxes? The war in the Middle East? Crab grass? Cholesterol? Nope, a builder like Lee thinks about an indelible pen that will mark a wet log. He thinks about a notch that can be cut with a jig to speed up production. He mentally argues the point where manufactured begins and hand-hewn ends. He dreams about a self-adjusting screw jack that will lower itself as the walls shrink and settle. He thinks about a faster way to level his log joists and a way to prefabricate his wall sections so he can squeeze out another house a year, because if he's good he's got more people who want houses than he'll ever be able to accommodate. And so he also thinks about starting a school and training more builders, figuring that the more good log houses that go up, the more popular they will ultimately be. He's probably right. Many of the rammed-earth and adobe builders I've talked to have the same theory about earth building. Like those types of houses, log houses are enjoying a comeback because serious builders like Lee stay awake nights figuring ways to combine the best of traditional techniques with modern production methods. And a few lucky families like the Densmores end up with houses that they can pass on to their great-grandchildren.

The area near the back door to the kitchen shows contrasting textures of log, siding, railings and stone.

The famous John Lee picnic table and benches are built out of log ends from the house construction.

An Introduction to Earth Houses

CHAPTER 9

Afficionados of earth-built homes know that earth structures are durable when stabilized and protected with a roof; that they are warm in winter and cool in summer; that they are termite-proof, rotproof and fireproof; that they can be built in just about any architectural style, and even be multistoried. Earth is used as a building material in all parts of the world, even wet and snowy areas such as New York State and Washington State. While earth houses are labor-intensive to build, they can be less expensive than conventional wood-frame structures, and can have excellent resale value. Most people have never seen an earth house first-hand, however, and few have been involved in a construction project using soil as a building material. Americans, except those in the Southwest, where the custom adobe house is now more expensive than its conventional competitors, usually associate earth-built houses with either third-world, underdeveloped technologies or with the early plains settlers. Words like "dirt" and "mud" don't inspire the confidence of the typical American consumer, builder or banker.

There are a number of factors responsible for this architectural narrow-mindedness. Because earth varies so much from location to location, there are no Universal Building Code standards, and thus building departments are leery of approving plans for earth buildings without a great deal of expensive engineering. No one can box and sell the material, so there are no lobbyists pushing earth construction. And since lending institutions are wary of unfamiliar, unconventional construction methods, the earth builder must first educate his or her banker before getting a loan.

The Church of the Holy Cross near Stateburg, South Carolina, is probably the best known earth building on the eastern seaboard. It was built in 1852 for a cost of $11,923.91. The rammed-earth walls are 22 in. thick and have about 2 in. of stucco and plaster. The gable ends are 43 ft. high. Photo by George Williamson.

Still, the situation for earth builders is far from hopeless. Building departments *are* granting permits for earth construction, and loans *are* being made for these types of houses. There is even a fledgling organization of earth builders striving for minimum standards and inclusion in state building codes. (New Mexico now has an earth section in its state code.) And as more earth houses are built—and more widely publicized—people are beginning to realize that an earth house that is built as carefully as a stick house with the same commonsense attention to drainage, foundation and roof will not wash away. (Indeed, some earth buildings have survived for hundreds of years without the benefit of roof or foundation.) Earth houses are not dirty; they can be stuccoed and painted, as well as left natural. They can have vinyl floors or tile floors or wall-to-wall carpeting. Generally, they don't look much different from their conventionally built counterparts, and in many parts of the United States uncommonly attractive earth houses are being custom-built. A recent ad in the real-estate section of a newspaper from Washington State's Puget Sound area lists a large rammed-earth island house for $300,000. An earth home in Stillwater, Oklahoma, that was built for $887.50 in the late 1930s sold a few years ago for $97,500. The current owners of the house have done some remodeling, and find that "this house is built like a bomb shelter. Even the inside walls are a foot thick. The earthen blocks can only be busted with a sledgehammer."

Types of earth construction

Earth building has taken different forms throughout history, but the types of earth construction can be broken into two groups: those using wet methods and those using dry methods. The wet methods are the cob, poured adobe, wattle and daub, and adobe block. The dry methods are pressed-earth block and rammed earth. I'll discuss each of these briefly, but of the wet methods I'll focus on adobe (used to build the house in Chapter 12), and of the dry methods I'll focus on rammed earth (used to build the houses in Chapters 10 and 11).

This beautiful plantation in South Carolina had approximately seven rammed-earth structures on it built between 1820 and 1850. The main house was originally built of wood, but the wings are built entirely of rammed earth. Photo by George Williamson.

Lyle Hymer-Thompson built this impressive adobe house in Redlands, California. The curved windows are Plexiglas. Photo by Lyle Hymer-Thompson.

The wet methods Of these, the easiest is the cob method. It's fairly primitive, and because no forms are used the walls are a bit wavy and uneven for modern building standards. To make a wall, the builder mixes up some wet clay and piles it on the ground until the pile starts to slump over, then the clay is allowed to dry. Standing on top of the wall, he trims the sides with a shovel to get the desired wall thickness. Sometimes straw, grass or other vegetative matter is added to bind the soil together and to keep it from cracking as it dries. Usually the base of the wall is much thicker than the top courses to support the weight of the walls and roof. Casa Grande in Casa Grande, Arizona, was apparently built with the cob method sometime in the 1200s. The main building is several stories high and is still standing, despite the fact that for several hundred years it had no roof nor received any maintenance.

Poured adobe, which is sometimes called puddled mud, is another wet method. To shape walls, wet soil and straw stabilized with hydrated lime are poured into some type of form. The material is stiff enough for the form to be moved as soon as the mix has been tamped in, but because it is too wet to support any weight, each short course must be allowed to dry thoroughly before the next course is added.

Michael Belshaw's system of poured-in-place adobe has forms that allow for curved as well as straight walls. All that is needed are the simple forms, a small mixer, a wheelbarrow and a couple of strong arms. The poured-adobe house above, built by Belshaw, has 36-in.-thick walls and the look of a traditional adobe. The buttresses are hollowed on top to carry off rainwater. Photo by Michael Belshaw.

The National Park Service built a protective roof over the unstabilized earth walls of Casa Grande in Casa Grande, Arizona, to prevent further deterioration of the structure. The walls were built by the cob method, also sometimes called coursed adobe in the Southwest, sometime in the 13th century. Since no forms are used in this method, it produces somewhat uneven walls.

This European-style wattle-and-daub wall is part of a building built like a 17th-century English barn at Renaissance Pleasure Faire in Novato, California.

A third method of wet construction, wattle and daub, dates back to the time of the Etruscans. It is probably one of the most common earth-construction techniques, and examples can still be found in Europe and the Mediterranean. In Australia, it was called "half-timber" construction; the Pueblo Indian version was called "jacal." Commonly used in a post-and-beam structure, wattle-and-daub walls are built on a screen of woven twigs and thin branches (the wattle). A mixture of mud and straw (the daub) is worked into the sticks, and then the wall is coated with a mud plaster. With a coat of whitewash, the walls look like they've been stuccoed. Wattle-and-daub walls are less durable than other earth walls, and need the protection of a long roof overhang. They also lack the insulative capabilities of a solid-earth wall.

The fourth wet method, and the best known, is the adobe block, which was introduced to North America by the Spaniards in the 16th century. Adobes were traditionally made from soil, water and straw poured into wooden forms on the ground and allowed to dry, but in the big adobe yards today, the proportions of clay and sand are carefully monitored and an asphalt emulsion is commonly added as a stabilizer. Adobes are still left to dry on the ground in the sun for several weeks and must be turned so that they dry uniformly. Generally, adobes are laid up with a mortar of about the same composition as the block. Today most adobe builders purchase their blocks from an adobe yard and have them delivered to the job site on large pallets dried and ready to use.

In a variation of the conventional method of adobe construction, adobe mud is blown onto laid-up adobe block with a concrete pump and modified gunite gun in a process called "ablobe." This technique can be used to add 6 in. of stabilized mud to an 8-in.-thick adobe block wall.

Adobes being laid up in a wall at an Earth Systems Expo '84 demonstration day in Albuquerque (left) are embedded in a mortar of about the same composition as the block.

The Santa Fe, New Mexico, rammed-earth home of architect Richard Yates, built by Yates and Donald Woodman, combines the traditional territorial adobe style with modern design. At left, a greenhouse separates the living room (foreground) from the children's bedrooms. Photo above by Donald Woodman.

The Church of the Holy Cross (photo, p. 94), built of rammed earth, is a National Historic Landmark.

These walls were rammed using commercially made Symons concrete forms. The forms use 4x4 panels and are braced with angle iron. Metal ties hold the forms together.

The dry methods Rammed earth and compressed-earth blocks are the earth-building techniques that use a dry-earth mix. Rammed earth has a long history. It was used in parts of the Great Wall of China. Hannibal reputedly introduced it to Europe in the third century B.C. In areas of France and Germany, some rammed-earth structures have been occupied continuously for over 400 years. Two of the oldest buildings in the United States, one in St. Augustine, Florida, and one in Santa Fe, New Mexico, are built of a type of rammed earth. The Church of the Holy Cross in Stateburg, South Carolina, was built in 1852 of rammed earth. It has been through an earthquake, a tornado, numerous hurricanes and over a hundred winters of 50 in. of rain. Its high, gabled walls and delicate arched windows are reminders that earth can stand as tall as any of its more traditional building-material cousins. In California there are some fine examples of rammed-earth construction built by the Chinese 150 years ago and still standing.

Most people are surprised by how dry the earth is in a rammed-earth mix: between 10% and 15% water by weight. For centuries, rammed-earth builders have followed the same basic technique and used a similar "ideal" soil mix of 30% clay to 70% sand. All you need to ram earth into walls are some forms the width of your wall and something to pound the soil into a compressed, rock-hard state. Early workers used forms about 3 ft. or 4 ft. long and about 2 ft.

high tied together with rope. As the soil was rammed with a heavy timber or a stick with a rock on the end, the forms were moved along the perimeter of the walls and then raised up for another course.

Today rammed-earth builders use forms made of plywood or metal, and fasten them together with steel ties or pipe clamps. Pneumatic tampers designed for foundry work or for backfilling are used to pound the earth into the forms. Most rammed-earth contractors also use some sort of front-loading tractor to deliver the material to the wall instead of the more traditional wheelbarrow and bucket. These builders would still prefer the ideal soil mix used by the old-timers, but by adding portland cement to the mix, they've found that they can successfully utilize many different types of soils. They've also proved that walls made with enough cement stabilizer may be exposed to the weather almost as safely as concrete.

One of the advantages of rammed-earth construction is that the walls are fairly solid even in their green state. The material does continue to harden over time, but there is no delay as the walls go up. Full-height walls can be rammed in a few hours. A good rammed-earth soil can be dug out of the ground and put directly into the wall forms, and the forms can be removed immediately. A crew of four or five with adequate machinery can quickly ram the monolithic walls for a small house in a day or two.

The basic unit of the forms shown here is a 3-ft. by 4-ft. aluminum panel designed for poured concrete. These are the forms used by Don Woodman and Richard Yates in their Santa Fe rammed-earth jobs.

The St. Thomas Moore Church of Margaret River, Australia, is built of stabilized rammed earth. Precast concrete arches were set on mortar beds on top of the rammed-earth piers. Completed in January 1983, the church cost $29,000 to build, or $420 a square meter. The ramming was completed in 60 days. Photo by Richard Woldendorp.

Closely related to rammed earth is the compressed-earth block. The advantage of compressed-earth block over adobe block is in the amount of moisture required to make one. Often the moisture that is present in the mined subsoil is sufficient to make good compressed block, but for an adobe block, about 22 gal. of water is needed for every cubic yard of material. To make compressed-earth blocks, the material can be tamped into small forms or pressed in a machine such as the Cinva Ram. (Currently in the Southwest there are several blockmakers who use hydraulic presses to turn out thousands of blocks a day—these would be ideal for use in areas short on water but long on suitable on-site material.) Compressed-earth blocks are laid in mud or cement mortar just as adobe blocks are. The blocks may be stabilized or not, as the builder prefers.

I've made compressed-earth blocks on my property with the Cinva Ram and even rebuilt a wall in my rammed-earth house with them. The blocks are pretty and strong, but I wouldn't recommend that anyone build an entire house with them. In our little project, we dug the soil out, mixed in portland cement, loaded the soil into the Ram, unloaded the block from the Ram, turned the blocks to cure (covered with plastic and sprayed with a hose), moved the blocks to the building site, and then laid them up in the wall using a mud-and-cement mortar. Looking back on the experience, it seems to me that no matter how organized you are, you have to handle the material for a compressed-earth block at least five times. That means some heavy lifting by the time you move the tons of material needed for even the smallest earth house.

World Wide Adobe's "Gold Brick 5000" is one of the many hydraulic pressed-block machines on the market. This one produces 600 blocks an hour. Soil is dumped into the hopper and blocks emerge on the conveyer belt.

At right is the Cinva Ram hand-powered block-making machine, with unstabilized blocks in the foreground. Optimists claim that 600 blocks a day can be manufactured with this machine. It was developed for use in third-world countries with little technical know-how but lots of available manpower. The author used the Cinva Ram to make compressed-earth blocks for one wall in his home, above.

Soils analysis

A basic knowledge of soils is essential to the earth builder. As a general rule, sandy clays and clayey sands make the best earth houses. But while a sandy-clay soil will make good rammed earth and adobe, a soil heavy in clay is best for adobe only. The problem with clayey soil is that it shrinks as it dries, and any wall that is monolithically built with it will be subject to serious cracking. If you use individual bricks of the same clayey soil, however, each brick will do its shrinking before it's laid into the wall. So a good rammed-earth soil will also make good adobes, but a soil that makes good adobes won't necessarily make good rammed earth unless it is stabilized (see p. 105).

Soils scientists rightly point out that the study of soils is complex and multifaceted. On the other hand, people having only a rudimentary knowledge of soils theory but a sound understanding of the behavior of their local soils have been building successfully with earth for years. With the basic observations and tests described in the following pages, I believe that even a novice builder can learn enough about the characteristics of his or her particular soil to predict accurately how it will behave in an earth-building project. If the commonsense approach fails, there is always the option of going to a soils laboratory for the information you lack.

Soils are classified by particle size. The largest particles are gravel and are anywhere from ¼ in. to 3 in. in diameter. Particles of sand are mid-size, about ¼ in. in diameter to as small as can be seen with the naked eye. Silt and clay are the finest particles—a typical clay particle might have a diameter of 0.00000004 in. (A pound of clay has 90 acres of surface area—no wonder it soaks up so much water.)

There are some general requirements for good adobe or rammed earth. As I said earlier, a soil with a ratio of 30% clay to 70% sand and gravel is ideal. The sand gives the soil its strength and the clay binds the coarser sand (and often gravel) together. A soil that is all clay will shrink and crack badly as it dries out, and reabsorb a tremendous amount of water when exposed to moisture. It's best to have a range of particle size—the various materials lock together better than if all components were of a similar size. Soils with a lot of decaying organic matter grow good vegetables but make terrible walls.

The earth builders I know approach the problem of soils testing in two different ways. One group finds a large source of a soil that seems suitable and then sends a sample to a soils lab for thorough analysis. The lab performs a series of tests, including the Atterberg Limits devised in 1911 by a Swedish soils scientist. These parameters are the liquid limit (the moisture content at which a soil becomes a liquid), the plastic limit (the moisture content at which a soil thread begins to crumble when rolled to a diameter of ⅛ in.), and the shrinkage limit (the minimum moisture content at 100% saturation). The lab will determine the plasticity index of the soil by figuring the difference between the liquid limit and the plastic limit. These tests tell the shape of the soil grains, the type and amount of clay contained in the sample, and the presence of organic components. The lab also does an unconfined-compression test, which measures the load a test sample prepared as it will be used in the field can withstand before it crumbles. A further test, the spray test, determines how well earth blocks or walls will hold up in a driving rain. A final evaluation that is especially helpful to rammed-earth builders is the compaction test. This determines the optimum moisture content at which the soil should be rammed. With a readily available soil that gets a good bill of health at the lab, builders in the first group use the same soil on all of their projects.

The other group of builders uses a series of field tests at each building site and chooses on-site or at least close-to-site materials that meet the minimum requirements for a good building soil. They use visual inspection and the mason-jar test to determine particle size. (All the tests mentioned here are further explained on p. 104.) Through a series of observations and tests, including the rolled-out-thread and ribbon tests, they can verify clay content. The drop-the-mud-ball test helps them determine optimum moisture content. Then they create small test blocks using whatever mode of earth construction they favor, and after the test blocks have dried, they observe their relative strength by abusing them—dropping them on a hard surface, striking them with a hammer and scratching at them with a knife blade. They soak the same test blocks in water or spray them with a garden hose to determine their resistance to erosion by rain. If the blocks fail, they try a different location for the soil or mix different soil components together. When all else fails, some builders add portland cement in percentage amounts by volume of compacted wall and make new test blocks, the idea being to add the minimum amount of cement required to make the soil strong and water-resistant.

The outside corners of the greenhouse of David Easton's home in Wilseyville, California, are made of 14-in.-thick rammed earth and are totally exposed to the weather. In the winter of 1982-83, they resisted 70 in. of rain as well as snow and freezing temperatures. Four years later, they show minimal wear on the outside surface.

Adobe builders are spared much of the worry about which soils will work and which won't because they usually buy their material from adobe yards. Most yards have the soils reports and unconfined-compression tests for their materials right there in the office—professional adobe builders won't buy from a yard that doesn't provide such information on request or that doesn't have a proven track record. A builder who is going to make adobes will need molds to pour the adobe mix into and lots of room to lay out the thousands of blocks necessary for an average-size house. The test procedures described here work well for adobe soils as well as rammed earth, but remember that a soil that is too clayey for rammed earth can still work for adobe.

I favor the do-it-yourself approach to testing, but I also realize that in some circumstances a laboratory report is necessary (such as when it is required for a building permit). My approach to earth building has been to become thoroughly familiar with the soils I deal with. Through my reading and contacts with soils experts, I now know some of the scientific principles behind what I have been doing, and I know the scientific names for the various clays and combinations of soils that turn up in the field. My real working knowledge, however, has come from simple observation, basic testing and actually building walls. From some simple projects at home I've found that the soil out by the barn is too clayey and needs more sand, but it will work the way it is with 12% portland cement. The soil up by the studio is full of rocks but just about perfect. The soil at the house is less clayey than out by the barn, but still needs sand—without the sand, it makes durable walls with 10% cement.

Whether you work through a lab or self-test your soil, remember to dig away the organic topsoil and take samples from the mineral soil underneath. Also, it's critical to label everything, so that when you find a suitable soil, you'll know exactly where on the site it came from.

Now I'll describe a typical testing procedure, which was used for a project I worked on in Mexico City in the summer of 1984, and which is also used by several of the earth builders I know. Together these tests double-check each other, and for reliable results all should be run on any soil sample under consideration.

A batch of nice-looking mineral soil was delivered to our Mexico site. The soil looked crumbly and not gummy—a good sign, meaning that there was a good range of particle sizes. The earth was still moist and compressed well in our hands when squeezed—another good sign. The first test we did was the mason-jar test. To perform this test, you fill a quart jar halfway with soil, then to the top with water. Shake the jar gently until all the soil is in suspension, then put it aside for an hour or two. When the water on top has cleared, you'll see several distinct layers of material. The coarse components (sand and gravel) will have settled to the bottom and the finer material (silt and clay) will be clearly defined in bands. In our case, it looked like the coarse material was about 4 in. in depth, compared to 2 in. of very fine material at the top—close to the ideal of 70% sand and gravel.

So far so good, but the next thing we needed to know was how much of the fine material in the top bands was actually clay. We were so confident that we had good soil that we skipped the thread test, which is the next step, and went directly to making our test blocks. We shouldn't have. We would have found that the sample had no clay, only worthless silt. We did find this out when the test blocks crumbled, but we could also have found it out from the thread test.

To do the thread test, you take an olive-size lump of soil and add enough water to make it pliable, not sticky. On a flat, clean surface (butcher paper on a board works well), use the palm of your hand to roll the lump back and forth until you make a thread. (It's the same way you probably made snakes out of modeling clay in kindergarten.) Try to roll out a thread ⅛ in. in diameter. If the soil has the right moisture content, the thread will begin to break into short segments as it reaches that point. If it breaks before it becomes that thin, add a little more water. If you can't get a thread going at any moisture content, you don't have any clay. As soon as the thread reaches the breaking stage, remold the sample into a ball and squeeze it between your thumb and forefinger. See how much pressure it takes to flatten the ball. If the remolded ball is really tough and doesn't crack or crumble when you pinch it, you've probably got too much clay (which would have been evidenced in the jar test by a very thin bottom layer of coarse material and a heavy top layer).

You can also put some wet soil on your hand and see how easy it is to wash off. Clayey soils will stick and take some rubbing. Silty soils run right off. Then you can do the ribbon test. Make a ribbon by rubbing a cigar-size lump of moist soil together between the palms of your hands. If you get a long ribbon going, you've got too much clay. If you can't get anything going, you don't have any clay. You can extrude the ribbon out over one edge of the bottom hand. The ribbon will droop, elongate and finally break. If the diameter of the ribbon is worked to about ¼ in. and breaks when it is about 1½ in. long, you have mechanical proof of a clay content between 15% and 20%. If the ribbon is longer than 1¾ in. at the breaking point, the clay content is 25% or higher. In Mexico, we would have found that we couldn't make a thread or a ribbon, and would have looked for some substitute material right away instead of waiting to see our test blocks fail.

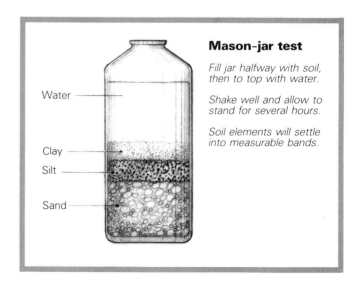

Mason-jar test

Fill jar halfway with soil, then to top with water.

Shake well and allow to stand for several hours.

Soil elements will settle into measurable bands.

Water

Clay

Silt

Sand

You can also taste and smell your soil. Unsuitable organic soils have a musty smell. I wasn't putting anything in my mouth in Mexico that didn't come out of a can or a bottle, but at home I'll bite a little soil. Sandy and silty soils are gritty and grind between your teeth. Clay soils are smooth and are like flour in your mouth.

The real proof of the pudding, however, is in the last test—the test block. If you can make a block that is tough, hard and won't dissolve or erode when you squirt a hose on it, you've got material that will work. For adobe, you can pour your mix into a can or small mold. For rammed earth, you can tamp the earth into a can or use a system such as David Easton, president of Rammed Earth Works, devised from some 2-in.-dia. PVC pipe. Easton takes a 6-in. length of pipe and cuts it in half lengthwise, then clamps the halves together with stainless-steel hose clamps. He tamps the soil sample into the pipe with a piece of ½-in.-dia. steel rod until it won't compact any further and then immediately unscrews the clamps and sets the test cylinder aside to cure. In Mexico, our samples crumbled as soon as we pulled off the clamps (fortunately, we were able to substitute a suitable source of good soil nearby).

Before ramming your test blocks, it's important to have the soil at the right moisture content, so you should do the squeeze-the-mud-ball test. Take a handful of soil and squeeze it together in your hand (about 3 in. in diameter). If it won't stick together, the soil is too dry. Hold the ball about 4 ft. above a hard surface and drop it. If the moisture is right, the ball will break apart into its original loose soil components. If it plops together like a mud pie, it is too wet. Ideal moisture is 15% by weight, and this test will put you close to that figure—the soil wants just enough moisture to make it stick together.

Most earth builders find they rely on the squeeze-the-mud-ball test in the field as well as when making test blocks. Ideal moisture gives optimum strength, and if you are way off, this test will show you right away. If there is too much moisture, the earth will stick to the forms and to the tamping tools; if too little, the earth will not stick together and the wall will crumble. Even if you have a lab report in your hand that tells you that 15% is the ideal moisture content for your soil, you'll have a devil of a time figuring that out as you add water with a hose to a pile of dirt whose volume and weight you have estimated. With the sun drying it out as you use it, and a hot wind blowing, there is no way to know the moisture content of your stockpile of earth—that is, unless you frequently use the squeeze-the-mud-ball test.

Additives

Earth builders talk a lot about binders, stabilizers and waterproofing agents. Some of these terms are used interchangeably and are confusing to the novice. All the terms refer to additives used to change the characteristics of the soil.

Straw is probably the most common binder. It helps a clayey brick to dry uniformly without cracking. Some builders still use straw in their clayey adobe soils.

A stabilized adobe block is not much stronger than an unstabilized one, but it is nearly waterproof and will not soak up water and soften like an unstabilized block. Stabilized rammed earth is much stronger than unstabilized rammed earth, but it still soaks up water—however, stabilized rammed earth will not soften when it gets wet. The reason the earth behaves so differently in these cases lies in the nature of the stabilizers used. Portland cement, the most common stabilizer for rammed earth, is a cementing agent, whereas asphalt emulsion, most common for adobe, is a waterproofing agent. (Hydrated lime is also sometimes used as a stabilizer with clayey soils—it actually breaks down the clay and creates cement-like compounds, but it takes much longer to cure than cement.)

Portland cement must be mixed dry with the soil and used immediately. This combination has traditionally been called soil cement. The cement works well in sandy soils, although some builders use it in all their rammed projects as extra insurance. U.S. Bureau of Standards tests conducted in 1940 show that the strength of a rammed-earth wall can be increased by as much as 500% with the addition of cement.

Asphalt emulsion is a liquid, and mixes easily with adobe mud. A typical stabilized adobe mud consists of one part cement, two parts soil (as close to the ideal of 30% clay and 70% sand as possible), three parts sand, and one and a half gallons of emulsified asphalt per sack of cement.

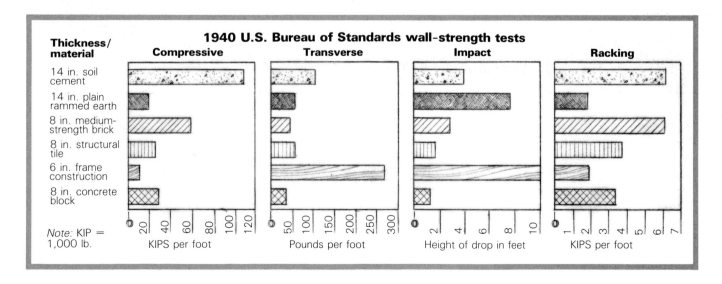

1940 U.S. Bureau of Standards wall-strength tests

Thickness/material:
- 14 in. soil cement
- 14 in. plain rammed earth
- 8 in. medium-strength brick
- 8 in. structural tile
- 6 in. frame construction
- 8 in. concrete block

Compressive — KIPS per foot (20, 40, 60, 80, 100, 120)
Transverse — Pounds per foot (50, 100, 150, 200, 250, 300)
Impact — Height of drop in feet (2, 4, 6, 8, 10)
Racking — KIPS per foot (1, 2, 3, 4, 5, 6, 7)

Note: KIP = 1,000 lb.

This house in Santa Fe, New Mexico, was built on speculation by the Rammed Earth Company under the direction of Donald Woodman and sold for a million dollars.

The best way to determine the optimum amount of stabilizer to use is to make test blocks with different percentages. A poor rammed-earth soil that has little or no clay can be made to work by adding cement. A very clayey soil can be made to work with a lot of cement or hydrated lime. (Lime works better with some clays than others. The best way to find out if it will work with a given soil is to make some test blocks without lime and then make some using 6% to 8% lime for comparison.) Builders who use portland cement add anywhere from 5% to 14%, depending on conditions. Here again, use test blocks to find what's right for your particular soil. If you need to describe the strength of your test blocks in numbers, you can make some testing equipment such as that described in the *Handbook for Building Homes of Earth* (see the bibliography) or you can take your test blocks to a lab. The Uniform Building Code requires a compressive strength of 300 lb. per sq. in. for an earth block—by adding cement, it's possible to obtain compressive strengths two or three times the minimum required.

When should you stabilize? If you've got a poor soil that won't work otherwise, stabilization is the answer. If you have to deal with adverse weather conditions and want the most protection for your walls, stabilization may give extra peace of mind. If you want to show the natural earth wall without any protective coating, you may want to stabilize. Why do some builders choose not to stabilize? They have ideal soil and feel they don't need extra strength or protec-

tion. They want to save the expense of the stabilizing agent and the extra labor and handling costs of mixing the agent into the soil. They plan to plaster the walls and/or use a long roof overhang to protect the walls from the weather. The reasons vary, but it usually boils down to economics. In California, David Easton uses a lot of different soils in his houses and figures that $200 or $300 worth of cement is cheap insurance against wall failures. In New Mexico, Donald Woodman figures that it's a waste of money with the ideal soils he uses. In the following chapters in this section, you'll see two houses that use stabilized earth and one that doesn't.

Some builders have speculated that additives can change the thermal characteristics of earth walls. Cement makes earth more dense, like concrete. Straw or wood shavings in a wet earth mix can add air pockets and increase insulative value. Some rammed-earth builders have experimented with leaving the center of the wall less compacted to increase insulative value, or even placing rigid insulation board in the middle of the wall and ramming around it. There is no research at this time to determine how effective these measures are. U.S. Government studies from the 1930s indicate that it takes 12 hours for heat or cold to penetrate a 14-in.-thick rammed-earth wall. This lag time, called the thermal flywheel effect, accounts for the fact that thick, high-mass walls store heat or cold for long periods of time. When outside temperatures fluctuate up and down, a thick-walled earth house tends to moderate the interior climate.

On the Mexico City project, we wanted to stabilize the walls on the west side of the house in case they were rained on before the house was plastered. Because we were mixing everything by hand, we didn't want to stabilize all the earth mix, so we decided to stabilize just the earth that went on the outside of the walls. This way only about one-fourth of the material had to be remixed with cement (the other soil was used just as it came from the ground). The only hard part for me was to learn the difference between *con cemento* and *sin cemento,* and to keep track of which bucket was which. (I also learned the Spanish for bucket, earth, water and tamper to fill out my basic Mexican building vocabulary.) We simply poured bucketfuls of stabilized soil up against the side of the form that was to be the outside of the wall. Ramming then proceeded as if each lift of soil was homogeneous. The triangular shape that the stabilized soil took in the form seems to have helped to key it into the rest of the wall. Doing the walls this way also allowed the less dense, unstabilized portion of the wall to provide more insulative value.

Requirements of the earth house

Assuming that the earth builder has a fundamental understanding of the materials, there are still some things he or she should be aware of before beginning to build. An earth wall can be constructed so that it holds a lot of weight, but not so that it can tolerate any lateral movement. In engineer's talk, earth is good in compression but poor in tension, so an earth building should be kept in a compressive mode. A wood wall will flex in a strong wind or earthquake; an earth wall will want to fall over. (Tall walls are more susceptible to these stresses than short walls.) Thus you must tie an earth wall securely to the other walls so that the whole building reacts as one unit to any lateral stress.

The bond beam is the most popular method used by earth builders to tie the walls of their buildings together. Bond beams can be made of either wood or reinforced poured concrete. Either way, the beams must tie solidly to the tops of all the earth walls and to each other. Picture the bond beam as a top-side foundation: it runs around the perimeter of the building at the top the way the foundation runs around the perimeter at the bottom.

Some builders go further in stiffening their earth walls. In California, which is earthquake country, some sort of vertical bracing is required within all earth and masonry walls. Lyle Hymer-Thompson (Chapter 12) was required to use vertical rebar in the adobe walls of his Redlands, California, project, and the same requirements are made in California for rammed-earth houses. Ironically, introducing rebar or other reinforcing material into rammed earth actually weakens the walls because the ramming process vibrates the reinforcing material and breaks its bond to the earth. So rammed-earth builders in earthquake country often structure their houses with vertical posts or poles tied into the foundation on the bottom and into the bond beam on the top. Generally, a sill is attached to the top of the bond beam and the roof system fastens to that. The earth walls, rammed in between the support posts and topped by the bond beam, are merely curtain walls. They fill in the space and keep out the weather

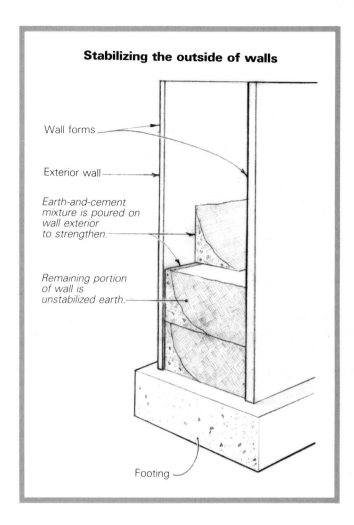

Stabilizing the outside of walls

Wall forms

Exterior wall

Earth-and-cement mixture is poured on wall exterior to strengthen.

Remaining portion of wall is unstabilized earth.

Footing

but don't do anything for the structure. If they were to fall down, melt or crumble, the roof would still stand.

My rammed-earth house (photos, pp. 108 and 109) has 8x8 pine posts on 6-ft. centers all mortised into an 8x8 wood bond beam. It is really a timber-frame house. David Easton still builds the same basic house, except that the posts and bond beams are made from poured concrete. With reinforcing steel through the foundation tied into vertical rebar in the posts, which is in turn tied into the rebar in the bond beam, this frame is very rigid. This method has also been used in California with adobe block—the adobes are simply laid up between the posts and topped by the bond beam.

The thickness of earth walls is largely a matter of personal preference. When building with adobe or pressed block, a builder is limited to the width of the block plus the thickness of insulation and/or plaster. Thicker adobe or block walls can be built by laying double block, but this increases labor and materials cost appreciably. For walls over 18 in. thick, rammed earth is the easier, more cost-effective method of construction. There are limits, however, to how high an earth wall can be built without buttresses (short walls that butt against the outside of the earth wall for support). The general rule is that height should not exceed 10 times wall thickness. Therefore, a 1-ft.-thick wall should not be higher than 10 ft., and a 3-ft.-thick wall could go 30 ft. high. The thicker the wall, the better the insulative value and the more thermal mass it provides.

Multistoried earth structures traditionally have had very thick walls on the bottom (2 ft. to 4 ft.) that taper to 1 ft. thick on the top floors. Tall earth buildings built of sun-baked mud are commonplace in the deserts of West Africa and Southeast Asia (see *Spectacular Vernacular* in the bibliography), and three- to five-storied rammed-earth buildings can still be seen in parts of France and Germany.

For all these heavy walls, earth houses need a reinforced-concrete perimeter foundation. The size of the footings will depend on the thickness and height of the walls and the nature of the soil you are building on. Local codes and conditions may require the services of a structural engineer. A look at the footing sizes for the earth houses in the following chapters will give an indication of what will be required. The *Handbook of Building Homes of Earth* (see the bibliography) gives some basic tests that the builder can perform to determine the strength of the soil the foundation will sit on and then recommends footing widths and thicknesses based on the number of stories, the thickness of the walls and the weight of the roof system. In cases where the soil is suitable for ramming and where the walls are not over 8 ft. high, a foundation equal in thickness to the earth wall and poured below the frostline is generally sufficient.

A common question asked by people who visit my earth home in the mountains is, "What do you do to finish the walls in an earth house?" There are lots of options. The easiest is to do nothing, to leave the walls as they came off the forms or as they look after the adobe blocks have been laid in place. Again, if you do choose to leave the earth in its natural state, you should stabilize it and/or cover it with a generous roof overhang. The exterior of the walls presents different problems than the interior. My walls are plastered on the outside with a dagga plaster discussed by Ken Kern in *The Owner-Built Home* (see the bibliography). It is the simplest plaster to use (next to a straight mud plaster, which must be renewed every season) and is made of two parts sand, one part finely screened soil and 10% plastic cement. It's basically a mud plaster with sand for strength. The idea is to use the same soil in the plaster as is in the walls so that you end up with the same coefficient of expansion. I've found dagga plaster works great as long as we don't let the sprinklers soak it and that it is easily repaired and patched if it cracks. Some people use a stucco plaster complete with wire base.

In *The Owner-Built Home,* Kern repeats some exotic recipes for homemade coatings taken from U.S. Dept. of Agriculture *Bulletins* published in the 1930s and 1940s. When you realize that most of these bulletins were aimed at American farmers, the ingredients aren't really as outlandish as they now seem. For example, a glue sizing can be made from six parts cottage cheese and one part quicklime. Buttermilk paint can be made from 4½ lb. of white cement to 1 gal. of buttermilk. Another coating formula is made from 30 lb. of flour and 50 gal. of water with some soil added after the flour and water have been cooked.

Many of the people I know have found that the wall interiors need to be painted or sealed. In my house, the walls were too dark, so my wife, Cathleen, painted them one at a time to lighten things up. She used an acrylic house paint with a primer directly on the earth, and it worked fine. Other builders have found that even with very hard walls, there is a certain amount of dusting that dirties the house—linseed oil or a polymer sealer will remedy this. Actually, you can put anything on earth walls that you would put on sheetrock walls—paint, plaster, wallpaper. You can even sheetrock over them.

The house above was built of rammed earth in Tucson, Arizona, by Randy Ewers and Terry Esser. Its lofty proportions challenge the preconception that rammed-earth buildings can't be multi-storied. Photo by Terry Esser.

The author's rammed-earth home (facing page), built by David Easton, has a sod roof sown with native pasture mix, wildflowers and red clover. At left, the interior floors are dyed and stamped concrete (for a discussion of this technique, see p. 118).

The House at Lake Camanche

CHAPTER 10

Jim and Frances Sturgeon saw their first rammed-earth house on television in 1981. It was one of David Easton's, the founder and president of Rammed Earth Works of Wilseyville, California. They couldn't wait to see a rammed-earth house in person, and when they saw one, they couldn't wait to have one. So they contracted Easton to build them a home on the shores of Lake Camanche, one of several man-made lakes in the rolling lowlands of the Sierra Nevada mountains in California. At the time, Easton had built only five rammed-earth houses, but they had garnered an inordinate amount of media attention and there was promise in the air. The Sturgeon house would prove to be pivotal in the development of Easton's construction system, and it would turn out to be one of his most impressive jobs.

Easton approached the design of the Sturgeons' house with a great deal of excitement. He was in love with their site and wanted to do something special with it. He envisioned a U-shaped house with the front door opening onto a hallway, or breezeway, leading to the patio. The breezeway would divide the house into two wings; each wing would open onto it through a weather door. At the far end of the breezeway would be another door, opposite the front door, in effect making the entire entryway a covered extension of the patio. With both doors open, the breezeway would frame a view of a little rock island in the center of the lake.

As the house plans evolved in discussions with the Sturgeons, Easton began to see the wings of the house as two separate units. He visualized the east wing being used for guests and visiting children—when not in use, it could be

closed to save heating and maintenance costs. The west wing would contain the primary living quarters: kitchen, family room, master bedroom, master bath and spa. But as it turned out, the Sturgeons use the whole house. Easton's intended sitting room in the east wing is now the living room and the intended bedroom in the southeast corner is the study/library. The bedroom in the northeast corner has been turned into a guest room. The Sturgeons eventually removed the doors to the wings, and now also use the breezeway as a room.

With the design done, Easton turned to testing the soil, and immediately found problems. For stone houses, the same basic types of stone will be found in most areas, and their hardness, porosity and durability can be easily identified. For log houses it's the same thing—there are a limited number of species of trees that make good log homes. For conventional stick houses, the materials are always the same—a 2x4 is a 2x4 is a 2x4. But building earth houses is more of an adventure. The earth isn't likely to be the same as on the last job, and indeed it may even vary from place to place on the site. The ingredients that go into each type of earth might be familiar, but the mixtures will probably not be. Even when you've had lots of experience, you can think that a particular soil is great, only to find it doesn't work at all. So testing, as described on pp. 103-105, is the key.

The soil on the Sturgeons' site was peculiar—white and powdery. It dried in the sun to a hard surface and then cracked like clay. The jar test revealed that it contained absolutely no coarse sand or gravel. In addition, round, water-polished cobblestones peppered the soil. So Easton made some test blocks from different mixtures of the soil and cement, with sand added to give the clay something to bind to. Eventually, he determined that the soil would work with 25% sand and 7% portland cement.

Most of the rammed-earth builders I know would never have fooled with this soil. They would have borne the small additional expense of hauling in a known good soil and then pounded it into a product whose quality and durability could be relied upon. Easton, however, prefers to work with native

Jim and Frances Sturgeon's rammed-earth home was one of builder David Easton's most ambitious projects. Nestled in the foothills of the Sierra Nevada mountains, the house features 3,000 sq. ft. of living space divided into two wings connected by a breezeway. The interior is enhanced by adobe-and-stone arched doorways.

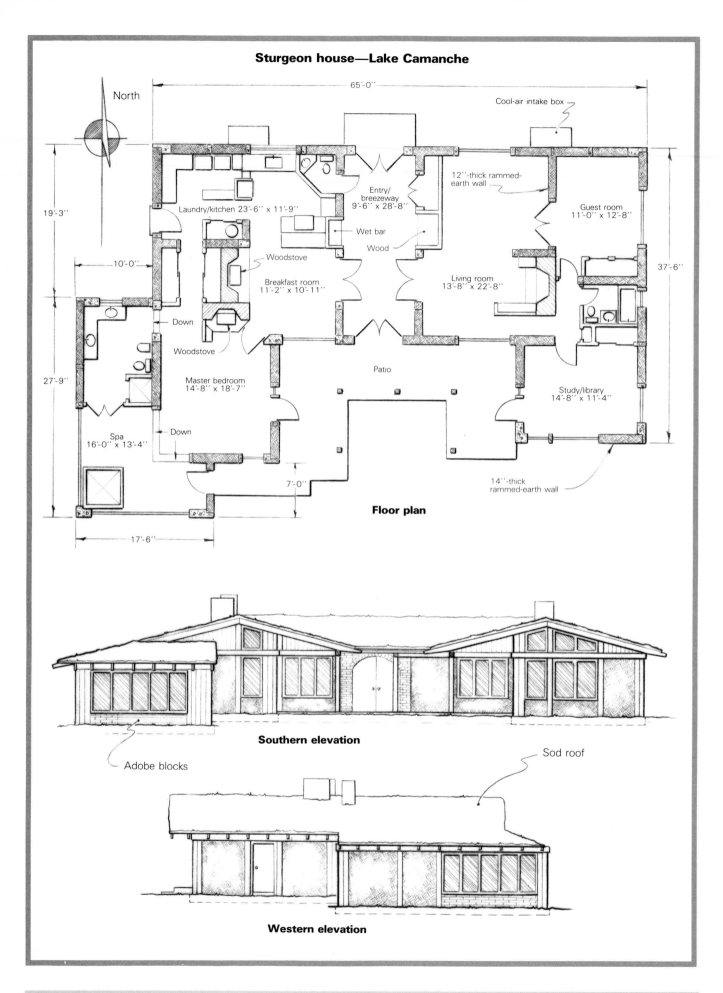

Sturgeon house—Lake Camanche

65'-0"

North

Cool-air intake box

19'-3"

Laundry/kitchen 23'-6" x 11'-9"

12"-thick rammed-earth wall

Entry/breezeway 9'-6" x 28'-8"

Guest room 11'-0" x 12'-8"

10'-0"

Woodstove

Wet bar

Wood

37'-6"

Breakfast room 11'-2" x 10'-11"

Living room 13'-8" x 22'-8"

Down

Woodstove

27'-9"

Master bedroom 14'-8" x 18'-7"

Patio

Study/library 14'-8" x 11'-4"

Down

Spa 16'-0" x 13'-4"

7'-0"

14"-thick rammed-earth wall

17'-6"

Floor plan

Southern elevation

Adobe blocks

Sod roof

Western elevation

soil, which he can do with a great deal of confidence because of his interlocking post-and-beam system of building—poured foundation, columns and bond beam are all linked structurally with steel to form a rigid, concrete box. The walls, as in any post-and-beam structure, merely fill in the spaces between the posts. They provide insulation and thermal mass, in addition to protection from the wind. They don't support the roof and need have only enough strength to keep standing, although after Easton had doctored the Sturgeons' soil, the walls could easily have supported the weight of the roof system and its heavy sod topping.

The Sturgeons' house was to sit on a little knoll, but because Easton prefers a level house pad for his projects, the first step in construction was peeling off the top of the knoll with a bulldozer. The topsoil was scraped off and stockpiled for later use on the sod roof, and the subsoil, or pure mineral soil, was stockpiled for the walls. Six inches of subsoil usually provides enough material to build all the exterior walls of a house; in this case, after the excavation there was enough for all the interior walls as well.

The foundation

At 3,000 sq. ft., the Sturgeon house had the largest and most complicated foundation that Easton had ever poured. It called for 30 yd. of concrete and close to 300 running ft. of foundation wall. Easton rams his walls in standard panels measuring 5½ ft. long by 6 ft. 10 in. high by 14 in. thick, each weighing about 3 tons—to support this tremendous weight, a footing below grade of 12 in. thick by 20 in. wide was required. (In this house, many of the interior earth walls are 12 in. thick, so footings had to be poured under those walls, too.) The stem wall extends 6 in. above grade, and is 14 in. wide at the top for exterior walls and 12 in. for interior walls. Creating a small ledge on both sides of the stem wall for the rammed-earth forms to rest on is a little trick that saves Easton a lot of time. (You simply nail a 1x2 to the top edge of the 2x6 stem-wall form.) The ledge is also a good place to rest the screed board when pouring the floor slab, and because it is level, it simplifies the process of plumbing the form boxes at that step in the building process (p. 117).

The foundation is reinforced with four pieces of #4 rebar (½-in.-dia.) laid horizontally: two pieces run along the bottom and two pieces along the top. Four lengths of vertical rebar are tied to the horizontal rebar at each corner and everywhere there will be a concrete column between earth wall panels. The pieces of vertical rebar are tied together from 3-ft. lengths and connect to the bond beam, which ties the tops of all the walls together. The result is a rigid unit—in effect a grade beam on the bottom joined to a grade beam on the top by a series of columns 6 ft. on center. After all the concrete has been poured, the earth walls are locked securely in place, giving them the shear strength to withstand the racking forces of a strong earthquake, and giving California building department officials the courage to sign off on a set of Easton plans.

Besides the ledges in the stem walls, another unusual feature of Easton's foundation system is the shutoffs he uses wherever there will be an exterior door. The shutoff is simply a piece of 2x6 blocking placed in the top of the foundation

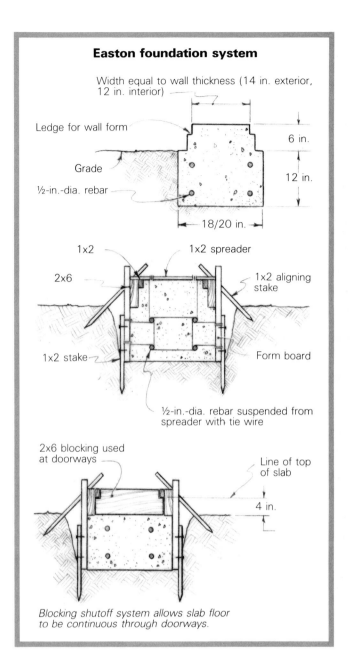

Easton foundation system

Width equal to wall thickness (14 in. exterior, 12 in. interior)

Ledge for wall form

Grade

½-in.-dia. rebar

6 in.

12 in.

18/20 in.

1x2

2x6

1x2 spreader

1x2 aligning stake

1x2 stake

Form board

½-in.-dia. rebar suspended from spreader with tie wire

2x6 blocking used at doorways

Line of top of slab

4 in.

Blocking shutoff system allows slab floor to be continuous through doorways.

Here the foundation is formed up and almost ready to pour. A cold-air intake box is visible in the foreground. The vertical rebar will tie into the rebar in the columns and corners.

formwork on either side of a doorway to drop the top of the foundation wall to 4 in. below the top of the finished slab. When the 4-in.-thick slab is poured, it can be run right through the doorways to a stoop or walkway. This also allows Easton to continuously stamp tiles (p. 118) into the slab through the doorways into the patio and entryway.

The Sturgeons' house is heated and cooled passively with a heating/cooling loop built into each wing. (Woodstoves in the breakfast room and master bedroom and an adobe fireplace in the living room help out in winter.) In the east wing, a pipe runs north to south under the floor from the guest bedroom to the study. Cool air from the guest bedroom is pulled through the pipe into the warm study, where it is heated and moved throughout the wing—there are no fans in the system, and the air moves naturally by convection. Cooling in summertime is aided by a lidded, concrete cool-air intake box on the outside of the guest-bedroom foundation wall; the box allows cool air to be brought into the loop from outside the building. (The lid is closed in the winter.) As the warm air rises, it is vented through skylights, high windows and greenhouse vents. In the west wing, the pipe runs from the north wall of the kitchen to the spa. (The cool-air intake box is outside the kitchen foundation wall.) The spa is below the level of the rest of the house, so the cool air spills into it beneath the wall to the master bedroom. The windows between the master bedroom and spa can be opened in winter to let the heated air flow back into the house and closed in summer so the house doesn't overheat.

The pipes under the floor in each wing are 12-in.-dia. culvert pipes and are dug in under the slab. Holes through the foundation for the cool-air intake boxes are made with pieces of 2-in.-dia. ABS pipe inserted in the forms before the pour—the small pipes allow adequate air intake without weakening the foundation.

On a complicated house like the Sturgeons', using a concrete pump with a boom and long hose to pour the foundation saves a lot of time. For a good bond with the earth walls, the top surface of the foundation is not troweled off but left with an irregular surface. These days, Easton pours the slab floor as soon as the foundation is done, but the Sturgeons' slab wasn't poured until after the walls were rammed. The advantage to pouring the slab before ramming is that you have a level area on which to drive the tractor while delivering the earth to the forms. This is especially helpful on a hilly site where the forms can't be conveniently reached from the outside. Another advantage is that the earth walls are easier to keep clean if the slab is poured, dyed and stamped before ramming—it was difficult on the Sturgeons' house to keep the concrete and the red concrete dye off the finished walls.

The lids of the two cool-air intake boxes are kept open in summer to allow cool air to be brought into the building from outside (above). Holes through the foundation made with pieces of pipe (right) pull outside air into the cooling loop.

Building the walls

The first steps in building the walls were to clear a mixing pad close to the house with the front-loading tractor, to ramp up dirt along the sides of the interior stem walls and to bend over the protruding rebar so the tractor could move freely within the perimeter of the building. Some of the wall forms would be filled from outside the building, but many would be more easily serviced from within. With the tractor, Easton put 2¼ yd. of soil onto the mixing pad, then added ¾ yd. of sand and 4 sacks of cement for a total of a little over 3 yd. of material. The pile was then rototilled to a uniform color. (The bucket on the tractor was used to fold the pile over and speed the mixing.) The rototiller operator picked through the pile and discarded large rocks and roots as he worked. By the time all the walls had been rammed, quite a pile of rocks had built up.

Moisture was added to the soil as needed with a hose attached to a gas-powered pump in the lake, and retained by a covering of black plastic over the pile. To monitor the moisture content of the earth, Easton used the squeeze-the-mud-ball test. Easton generally prefers a moisture content of 12% to 15% by weight.

The problem with ramming earth on a 20th-century-production basis has always been that it is labor-intensive. Traditional slip-forming is far too slow—a 2-ft.-long or 3-ft.-long form that could do 18 in. or 24 in. of wall height at a time had to be disassembled, reassembled, plumbed and fastened so often that a small crew spent as much time fussing with the forms as they did ramming earth. Earth-house contractors discovered early on the need for forms that once set up could be used to ram large portions of wall. One solution is to form the entire house at one time, much as in a large concrete pour, and then maybe reset the forms once more for added wall height. This requires a sizable investment in forms and vehicles for transporting them, but once

the forms are set, the ramming crew can pound walls without stopping.

Because Easton came to the field as an owner-builder, he wanted a method that was quick and efficient but also reasonably priced. He wanted his forms to be light enough to be portable and easy to assemble and disassemble, yet sturdy enough to withstand the punishment of power tampers and tractor buckets. (An initial design made of steel with pipes and cables for cranking the forms up the walls lies rusting in his boneyard. It promised to be extremely durable, but was much too heavy to operate.) Easton also wanted walls without holes from form ties, so that he wouldn't have to plaster them. These considerations led him to develop a modular building system—each house is constructed of a series of large, freestanding earth panels with poured-concrete columns in between. The pipes that hold the forms together run in the spaces between the panels and therefore the earth remains unmarred.

Easton's forms have evolved over the years, but all the variations have remained true to his original goals. Basically, he erects a box with plywood sides and end boards on the foundation, plumbs it, rams it full of earth, pulls the form off and moves it to the next spot. The box is stoutly braced with 2x boards; pipes and pipe clamps squeeze the sides together and hold the end boards in place. The modular system is versatile—he can use the same set of forms for shorter wall sections simply by setting the end boards closer together. Easton has also built walls as high as 17 ft. by stacking the plywood boxes on top of each other. Only two or three forms per house are needed, for once the earth is rammed, the forms can be removed immediately and set up in a new location. The freshly rammed earth will harden for months, but even though the surface and corners are fragile, the earth will immediately support its own weight and even the weight of two or three workers standing on top.

Rototiller operator Ted Toren mixes cement and sand into a pile of mineral soil for the form in the background.

Above left, Duncan Houghton rams an earth panel with Easton's old-style forms. Above right, the crew clamps together an early form box. The side that would be on the inside of the house was made from a full-height piece of plywood, to avoid a seam in the finished wall. The other side was made of two sections of plywood. Bracing nailed directly to the plywood made these forms extremely heavy.

Here Jamie Scott rams a full-height panel with the new forms Easton has developed. The cleats on the ends of the 2x10 whalers hold them over the pipe clamps.

When I joined Easton's construction company in the late 1970s, he was using the forms with which we started the Sturgeon house. They were made of 1⅛-in.-thick plywood with 2x8 bracing nailed on. Walls 7 ft. high were made by stacking a 3-ft.-high form on top of a 4-ft.-high form. There were 24 ft. of 2x8s and 8 ft. of 2x4s nailed to each side of a 4-ft.-high form. Those 2x8s came from Easton's mill, and when he mills 2x8s they actually measure 2x8. When Easton built a new set of forms out of green 2x8s right off the mill, they were technically portable, but it was impossible for one worker to handle them alone.

At this time Easton was also using a form having one 7-ft.-high side and the other made of a 4-ft.-high and a 3-ft.-high section. Each form consisted of five pieces of wood: the full-height side, the two half-sides, and the two 7-ft.-high by 14-in.-wide end boards. By facing the full-height side with tightly spliced sheets of Masonite, Easton could produce a wall that was seamless on one side, important in the days when he didn't plaster. It was a good idea, but it really pushed the concept of portability to the limit, because with the 2x8 bracing nailed on, the full-height side weighed much more than the already heavy half-sides. The form was hard to set up, too. The full-height side had to be held by two people while the end boards and first half-side were put in place. Once the first half-wall was rammed, the top half-side was put on and ramming completed. To disassemble, the whole crew was needed—a big difference from Easton's current form, which can be assembled by one worker and easily disassembled by two.

With a job the size of the Sturgeons' house, Easton figured that he needed better forms or he'd be bogged down building walls forever. Somewhere in the middle of building the garage Easton had a brainstorm, and the forms he developed then are basically the ones he uses today. The forms are still plywood boxes, but the bracing is not nailed on, making them much easier to handle. The bracing, made from 8-ft.-long 2x10s, called whalers, rests on pipes and is cinched up to the plywood with Pony clamps. (The whalers also make a handy ladder for climbing up the forms and a good platform for the tamper operator to stand on.) The most anyone has to lift at one time is a sheet of plywood or a single 2x10.

Easton makes the forms from sheets of 1⅛-in.-thick tongue-and-groove plywood, 42-in. lengths of ¾-in.-dia. black steel pipe, Jorgensen Pony pipe clamps and enough additional plywood to make end boards the width of the walls. (Tongue-and-groove plywood makes it possible to lock the top form box onto the bottom one, creating a less visible line at the joint.) The bottom sets of forms use 4x8 sheets of plywood, and the 1¼-in.-dia. holes for the pipes are drilled 12 in. in from each end of the plywood at 3 in., 17 in., 33 in. and 46 in. from the bottom edge of the plywood. The top sets of forms use 3x8 sheets of plywood and the holes are drilled at 3 in., 17 in. and 34 in. from the bottom edge. If you are building 12-in.-thick walls, the end boards can be made from the 8-ft. by 12-in. strips that are cut off to make the 3x8 sheets of plywood. On some jobs, Easton nails beveled 2x6s onto the inside of the end boards

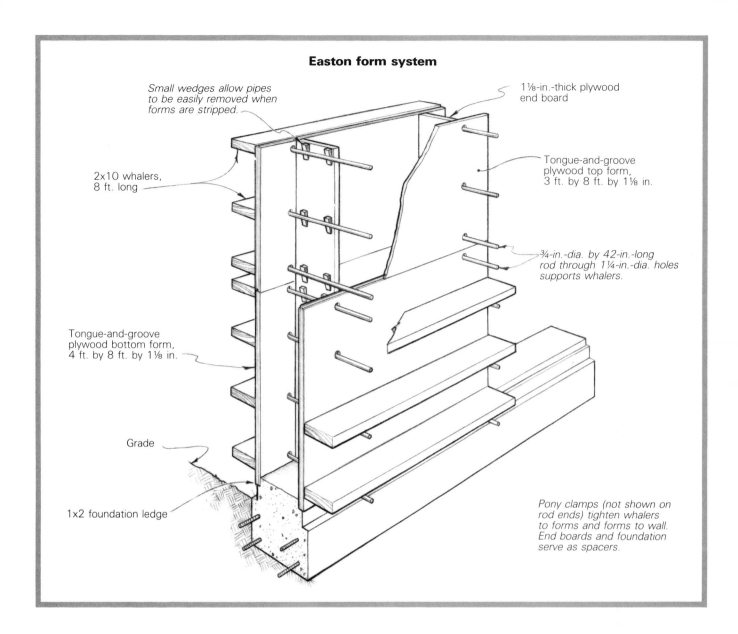

Easton form system

Small wedges allow pipes to be easily removed when forms are stripped.

1⅛-in.-thick plywood end board

2x10 whalers, 8 ft. long

Tongue-and-groove plywood top form, 3 ft. by 8 ft. by 1⅛ in.

¾-in.-dia. by 42-in.-long rod through 1¼-in.-dia. holes supports whalers.

Tongue-and-groove plywood bottom form, 4 ft. by 8 ft. by 1⅛ in.

Grade

1x2 foundation ledge

Pony clamps (not shown on rod ends) tighten whalers to forms and forms to wall. End boards and foundation serve as spacers.

to make a keyway, but the beveled board is tricky to remove without taking a hunk of wall with it. Thin walls are especially fragile. When he doesn't use the keyway board, Easton digs a keyway into the wall with a claw hammer as soon as the forms are stripped off the wall. The form as described produces a standard earth wall panel that is 5½ ft. long.

Assembling the forms is simple. Two sheets of 4x8 plywood are put on the foundation ledge and held in place temporarily with a pipe clamp tightened loosely across the top. The end boards are slipped in, and the pipes run through the holes. When the first couple of pipes are in, the temporary clamp is removed, and when all the pipes are in, the whalers are laid in place. The frogs are slipped over the ends of the pipes and pushed up against the whalers. The box is checked in all directions for plumb before the clamps are tightened, and wooden wedges are inserted between the pipes and the end boards. Before stripping off the forms, the workers knock out the wedges and free the pipes for easy removal—without the wedges, the end boards jam against the pipes from the pressures of ramming. As Easton says, you forget to put the wedges in only once.

The tractor dumps a load of soil into the form box.

So that he wouldn't have to shut down the tractor or the tamper to set up, Easton built three sets of standard forms and a couple of narrower ones that would produce walls 4 ft. and 3 ft. long. Once the crew got going, they were able to keep the tamper busy most of the time. One worker would set up the next set of forms while the last one was still being filled. Once the form was topped off with earth, the worker on the tamper moved to the next form and two workers dismantled the completed one. The latter was then reset at another point on the foundation before the worker on the tamper was ready to move on.

Enough soil is delivered to the forms by the tractor to create a 6-in.-deep layer—the worker on the form spreads it with a shovel and kicks it around to make it more or less even. The soil is then compacted with the backfill tamper to a bit less than 4 in. The tamper is hooked to a 100-cfm compressor with the air pressure set at 80 psi. A pneumatic tamper like this with an 8-in.-dia. pod on the end will ram walls just about ten times faster than a hand tamper. The ramming of a standard 3-ton wall panel takes about 45 minutes, including the time it takes to stack the second form box. The ramming of the entire Sturgeon house and garage took about nine days.

As the walls were completed, they were wrapped in black polyethylene to keep them from drying out too fast in the increasingly hotter days of late spring. Rammed earth cures slowly, and in hot, dry weather the surface of the walls will crack if they are not kept moist. Most of these curing bags were left on for a couple of weeks to give the walls a good cure. They also helped to protect the walls from being spotted with red-dyed concrete during the next step of construction, the pouring and stamping of the slab floor.

Pouring the floor: instant tiles

Concrete finishers have been dying concrete and stamping patterns into it for years—it's an efficient, inexpensive way to get a good-looking floor. You can get stamps for tile patterns, brick patterns, cobblestones and even flagstone patterns. The tools for stamping look like giant cookie cutters and can be purchased from concrete supply houses across the country, as can be the dye. The normal procedure is to dye the concrete with one of the many colors available and then to finish-trowel the slab before stamping the pattern. Easton is now experimenting with soil-cement stamped floors, but he used concrete in the Sturgeons' house.

The 4-in.-thick concrete slabs in the house and garage were prepared in the normal way with a plastic vapor barrier, a 2-in.-thick layer of sand and a layer of wire mesh. The pour was divided into several areas to give time for the finishing, dying and stamping. Red Lithochrome dye was first mixed with fine sand, then the mixture was broadcast over the wet, troweled concrete and worked in with the floats. The slab was then finish-troweled and a guideline snapped into the wet mud. Each 2-ft. by 3-ft. stamping tool for the 12-in.-square tile pattern chosen by the Sturgeons was pushed (or pounded, if the mud was setting up) in along the line. The tools were designed to butt together to keep the lines square. Once a tool was embedded, the worker would stand on it to stamp the next block, and so on until the last tool was used. Then the first tool would be pulled out and leapfrogged to the front of the line. At the edges of the wall or around fixtures, lines were made for partial tiles with a small hand tool that looks like a large broad-faced chisel. After the stamped slab had dried, the surface was waxed and then grouted with gray tile grout.

At right, Easton (standing) and George Clark set the stamping tools on the east side of the patio area. The poured and stamped slab in the garage, far right, is ready for grouting.

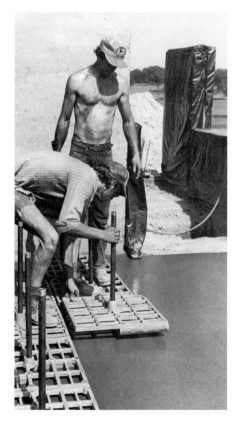

The beam-and-column system

At this point in the construction, the freestanding, rammed-earth wall panels were all up and the stamped slab had cured. Each panel was separated from the next by a 6-in. space, and in the evening sun, the huge wall blocks looked like a modern interpretation of Stonehenge. The next step was to tie all the panels together with poured-concrete columns and a bond beam. In addition to connecting the walls, in Easton houses the bond beam, faced with pieces of 5/4 cedar, is used as a header for the doors and windows—the 7-ft. height of the walls allows a manufactured door system to be framed under the header. Easton also uses 5/4 x 8 cedar to form many of the columns, leaving the wood in place after the pour to trim out the earth panels. (The bond beam is formed with 2x8s.) To create the 14-in.-wide boards necessary to match the thickness of the walls for window and doorway trim on the Sturgeon house, Easton cleated a 6-in.-wide piece of 5/4 cedar to an 8-in.-wide piece. Cleated boards are also necessary for the underside of the bond beam where it crosses a door or window opening.

Before building the formwork for the columns and bond beam, Easton ties four pieces of ½-in.-dia. rebar from the foundation to the columns. These four pieces are in turn tied to the two lengths of rebar that go around the tops of the walls inside the bond beam. Then a horizontal ribbon of 5/4 cedar is run along the top of the wall panels to form the bond beam. Vertical pieces of 5/4 cedar are used to cover the spaces between the panels (the columns). The cedar overlaps the edges of the earth walls by 1 in., and is held in place by ¼-in.-dia. threaded rod that is tightened with nuts set in countersunk holes. The form wood for the bond beam is held together by a combination of threaded rod and 1x2 wood spacers. After the concrete is poured into the columns and bond beam, the end of the threaded rod is snapped off and the holes plugged with wood dowels.

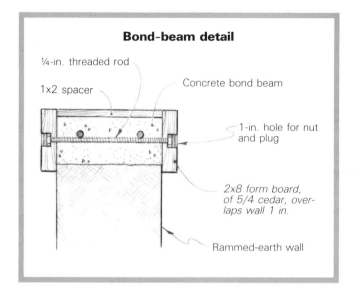

Bond-beam detail

¼-in. threaded rod

1x2 spacer

Concrete bond beam

1-in. hole for nut and plug

2x8 form board, of 5/4 cedar, overlaps wall 1 in.

Rammed-earth wall

This is a bird's-eye view of the bond beam looking toward the east. The adobe-and-stone archway in the foreground leads from the master bedroom to the kitchen area.

Nowadays Easton rams his corners out of earth, but at the time he was building the Sturgeons' house, his usual procedure was to run rebar up from the foundation to the bond beam on each corner and form what looked like gigantic columns. He used 5/4 cedar for these corners, and also usually made a concrete-filled wood column at each side of a doorway by building two three-sided boxes out of the same cedar. Easton had already decided to use adobe block instead of cedar on some of the columns and corners of the Sturgeons' house, and had also designed a series of arches for the doorways, a fireplace in the living room and huge hearths for the woodstoves, all from adobe. But as he watched the pile of rocks grow during the construction of the walls, he decided to incorporate the stones into the corners, columns and arches. The arches were formed up in much the same way as those discussed in Chapter 4: Easton cut two pieces of 1⅛-in.-thick plywood (from an old set of forms) into a half-circle, then nailed the plywood to 2x6 framing members to end up with a wooden arch form that was about 8 in. thick. He braced each form into the doorway with 2x4 posts, and laid the adobes and stones right on top of them. The adobe-and-stone-trimmed columns, corners and arches contained the same rebar used throughout the house, and they were poured full of concrete at the same time the wood-trimmed members were poured. The arches then became part of the continuous bond beam.

The poured columns are good, convenient places to hide electrical runs and plumbing vents. Easton attached electrical boxes to the inside of the form boards and ran electrical conduit up the board and over the top of the bond beam. He covered the end of the conduit so that concrete wouldn't get in during the pour. In the laid-up columns, Easton mortared the electrical boxes right in with the adobe and stone, then ran the conduit to the top of the bond beam—once the columns and bond beam are poured, it is an easy matter to thread Romex wiring along the top of the bond beam and down through the protruding conduit to the electrical boxes. Vent pipes are also easily run up the columns before the concrete is poured.

Once all the formwork was in place and the stone and adobe work completed, Easton did the pour. He used a concrete grout mix (¼-in. aggregate) for this and did it in two passes so as not to put too much pressure on the columns and blow out the forms. (We had blown out an entire concrete corner on the previous house, and we were all paranoid about doing it again.) While the bond beam was still wet, J-bolts were set every 2 ft. down the center on top. Later, these bolts would secure the 2x6 top plate.

This view is from the patio area through the entryway. Note the electrical boxes in the ends of the bond-beam form wood for the patio lights.

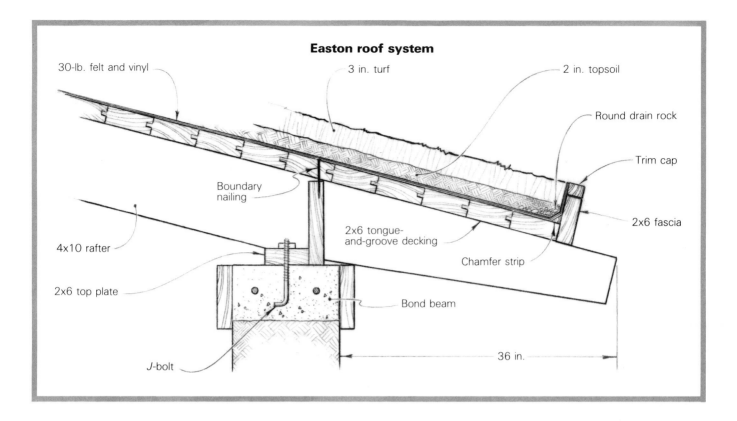

Easton roof system

30-lb. felt and vinyl

3 in. turf

2 in. topsoil

Round drain rock

Trim cap

2x6 fascia

Boundary nailing

2x6 tongue-and-groove decking

Chamfer strip

4x10 rafter

2x6 top plate

Bond beam

J-bolt

36 in.

The roof

With the bond beam and columns poured, it was time to put on the roof. A 2x6 top plate was set on the top of the bond beam and bolted down to the J-bolts that had been placed in the wet mud of the beam. The lumber for the ridge beams and rafters was all milled out of fir and ponderosa pine at Easton's mill: 4x10 rafters are notched into the 6x12 ridge beam, and rebar pins are drilled through the rafter ends and into the beam. The roof was built up on 2x6 tongue-and-groove pine left exposed on the underside. A layer of 30-lb. felt covers the boards and a 20-mil vinyl membrane covers the felt. The membrane is glued up with 4-in. seams and folded and glued around all the skylights. The topping for the roof is 3 in. of turf rolled out over 2 in. of enriched topsoil. The growing medium for the sod was made up of sawdust from the mill, turkey manure, and the topsoil scraped from the site during the initial excavation for the house. A sprinkler system covers the entire roof area and is automatically controlled to cool the roof in summer heat.

Laying the sod roof was a big event on this job—even a San Francisco television crew from *Evening Magazine* came to film this part of the construction as an introduction to Easton's houses. The rolled sod, which had been delivered by flatbed truck, was hauled onto the roof by a volunteer crew of seven, who shone with clean faces, clean shirts, combed hair and ready smiles. Easton had installed a set of ramps to walk the sod up, and the eager crew had the whole roof covered with grass in little more than two hours. The roof works well in the summer in this arid area, cooling the house like a giant evaporative cooler. In the Lake Camanche area, winter temperatures seldom drop below freezing, and the thatch of roots and grass that grows long over the summer provides a certain amount of insulation.

Here the rafters are on and the rafter blocking is in. The chimney flue liners are in, but the stone-and-adobe facing has not yet been applied.

The north wall shows earth walls, columns faced with stone and adobe, a cool-air intake box, and the grillwork that the Sturgeons applied to the windows.

The living room in the east wing has a fireplace made of adobe blocks (above). The kitchen of the Sturgeons' house (right) features a dyed, stamped and grouted concrete slab, and an arch laid up of adobe and stone.

With the roof on, the inside spaces began to look like a house. The views through the rooms with the repeating arches of adobe and rock were inviting. The cabinets and cupboards, the trim, and the wood above the earth walls were all made from native pine and cedar milled at Easton's mill. The Sturgeons bought ceramic sinks and tile in Mexico for the bathrooms. Doors were built from tongue-and-groove pine edge-glued and hung with handmade hinge straps and hardware. The double doors at both ends of the entryway are quite handsome. Windows were also made on site from Easton's ponderosa pine. The Sturgeons added iron grillwork to the windows and put terra-cotta clay hangings on the outside walls. More of the round stones were used to fill the drain line that runs along the eaves and also to build planter boxes along the length of the north wall. A lovely little pergola sets off the entryway, and with swings on both sides makes a comfortable place to sit on a summer afternoon.

Easton had left all the walls unfinished except for a sealer coat of linseed oil. A year later, Frances Sturgeon decided to have the inside of the house plastered to prevent the walls from dusting and to lighten their color. Two years ago, an incredibly wet winter brought some fierce storms and driving rains that beat on the south-facing walls for weeks. The wettest walls soaked nearly through, and in spite of their stabilization began to soften on the surface. As soon as he was able, Easton plastered those walls, too, and today they show no apparent damage.

The long 36-in. overhang was supposed to protect the walls from prolonged moisture, but there is not much to do to protect against driving, horizontal rains. In his early houses, Easton tried especially hard to get walls that looked good straight off the forms. Most of his clients have liked the look and feel of the uncovered and uncoated earth, but after the Sturgeon experience, Easton is encouraging people to plaster. He figures that it's cheap insurance and the money spent on plastering is probably saved in building walls that don't have to be cosmetically perfect. For one thing, when Easton plasters, he doesn't have to worry about the seam between the two forms on full-height walls.

The Sturgeons' house is a good example of what Buckminster Fuller meant when he said that a house should be trimmed and steered like a sailboat. There are no mechanical systems for heating and cooling, but the house works well with some steering by its owners. The skylight shutters (designed by Easton) are opened at night in the summer and in the daytime in cool weather. The airtight woodstoves are tended during the winter, and in the summer the sprinkler system is programmed for maximum cooling. The cool-air intake boxes are opened in summer and closed in winter.

Easton went back to the Sturgeons' house a year after it was completed to add a covered carport to connect the house and the garage, and to change the latter into an apartment and guest cottage. Like the other two earth buildings, the carport has a sod roof.

On the pergola outside the front entrance, swinging benches make a comfortable place to sit on a summer evening. The piers for the pergola posts are laid up with adobes and stones from the site.

The House in Estancia Valley

CHAPTER 11

After 42 years of building roads in New Mexico, Don Huston was ready to retire and start something new. His son, Stan, had worked his way up through the building ranks to a contractor's license and was building houses in the Albuquerque area. He, too, was ready for a new adventure. When they both read an article in *Popular Science* in 1981 describing David Easton's rammed-earth system, they decided that rammed earth was the way to go. It took until 1983 to get organized, but by then the newly born Huston Construction Company had its first rammed-earth project, a retirement home for Don and Mary Huston.

In the days when the Huston family was moving around the state from road job to road job, they had lived in several adobe houses, and Don Huston had then dreamed of building an adobe house of his own design. When the Hustons bought their 420 acres of wind-swept grama grass east of Albuquerque in 1963, there was a 50-year-old adobe house on the property. Huston figured that local soil had been used to make the adobes, and further assumed that the area would make a good adobe yard. He got permission from the county to build a yard, and designed a mechanized mold and a machine to lay adobes quickly and efficiently with his son. They had gone so far as to contract a neighbor to start building the molds when they had a major change in plans.

Huston particularly liked the idea of the homogeneous, monolithic walls without mortar lines that ramming earth allowed, and the efficiency of the system—he wouldn't have the endless handling of material that adobe-building entailed. Soon he had tracked down all the old, hard-to-find

classics on rammed earth (see the bibliography), but he got the most help from visiting Tom Schmidt, a long-time rammed-earth builder in St. David, Arizona. Schmidt was building a two-story house with 3-ft.-thick walls, and his partners, Quentin Branch and Bill Knauss, were building a house with 2-ft.-thick walls covered on the outside with 2 in. of expanded polystyrene insulation. Huston couldn't wait to get home to start designing a house and some rammed-earth forms of his own.

At the 7,000-ft. level in Estancia Valley, winter temperatures drop below 0°F and summer temperatures exceed 95°F. Huston wanted a house that would be comfortable in all seasons with as little supplemental heating and cooling as possible. To buffer against the extremes of temperature, he designed the exterior walls to be more than 3 ft. thick (34-in. walls with 2 in. of insulation plus plaster) and the interior walls 18 in. thick (16 in. of earth plus plaster). To Huston, two-thirds of the charm of an earth building is the thick walls, and he feels that the difference between a 2-ft.-thick adobe or rammed-earth wall and one that is 3 ft. thick really doesn't affect the performance of the house (especially when you insulate the outside of the walls as he does). Because he calculated that the walls represented less than 20% of his total cost, a thicker wall was not going to break the bank.

Huston packaged his house with a traditional New Mexico pueblo look—exposed pine vigas, rounded corners and brown plaster. He oriented it to the south, and built the south wall from 2x6s and lots of glass to take advantage of the warming New Mexico sun. For the garage, Huston called for 2-ft.-thick walls and roll-up glass garage doors. He figured that the solar gain and the mass of the walls would combine to keep the cars warm enough to start easily. (Even with outside temperatures down in the single digits, the temperature in the unheated garage has never dropped below 45°F.) During the winter of 1984-85, when snow was on the ground from October to December and temperatures dropped to −8°F, the Hustons' natural-gas heating bills were $70 to $80 a month. (There's a 35,000-BTU gas

The Huston family built their retirement home on 420 acres east of Albuquerque, New Mexico. With 34-in.-thick rammed-earth walls, plenty of insulation, and an expanse of glass on the south side, the house is comfortable year round, despite fluctuations in outdoor temperatures. Shown here is the entry foyer on the east side of the house.

Exposed vigas, rounded corners and a flat roof with parapet are all elements in the pueblo style of southwestern architecture. The concrete apron slopes away from the house and keeps water away from the rammed-earth walls.

The garage has glass roll-up doors that let in enough sun in the severe winter to permit easy starting of the family truck and cars.

panel ray heater in each wing of the house for supplemental heating.) They used no cooling in the summer and kept inside temperatures in the comfortable mid-70s. Fortunately, once Huston had worked out the plans for the house, he was able to work with a knowledgeable building inspector (unlike most states, New Mexico has an earth section in its code), so he didn't have to defend his design.

The Hustons' house is beautiful and substantial. I particularly like the deep window seats made possible by the extra-thick walls. Huston also used the wall thickness to create an airlock entry to his office, simply by hanging a storm door on the outside of the wall and a wood door on the inside—there's just enough room to close the outer door before opening the inner. The inside of the house is plastered with white gypsum plaster, and is light and airy throughout. The generous use of glass on the south side, several well-placed skylights, glass doors and clerestories on both sides of the 12-ft.-high living room bring light into the building. The light-colored walls contrast nicely with the log roof beams and the exposed underside of the roof decking. Unpainted interior wood doors, old barn-wood lintels and oak cabinetry enhance the natural look. Except for the niches in the living-room walls that hold the Hustons' collection of Indian art, the interior of the house is fairly conventional. Carpeting, comfortable furnishings, tile countertops and large rooms create an inviting environment. The Hustons' home proves that an earth house need not be rustic or trendy.

Windows placed to the outside of the walls leave nearly a 30-in.-deep window seat. Interior walls are unpainted gypsum plaster.

The living room bisects the house into two separate living areas. Most of the clerestory windows are fixed, but some are operable.

This south-facing side of the house, nearly all glass with framed and plastered walls in between, contributes to the light and airy interior. Solar panels cover the sloped end of the living-room roof.

Site preparation and the foundation

To level the site and build up the pad just enough so that water would drain away from the house on all sides, Huston brought in an elevating scraper. These behemoths look more at home working on a new freeway than on a house site, but Huston also used his to mine the subsoil for the walls from a spot he found just south of the house. This soil was built up into a large pile, then Huston placed a sprinkler on top to soak the earth. As it dried, it would be used as it reached optimum moisture content. In less dry climates, builders often find that the subsoil has the optimum moisture content just as it is dug from the ground, but in Huston's part of New Mexico, even the subsoil comes up bone dry.

The Hustons realized that they would be faced with different soils with each new job, so they decided to experiment a little on their house. (In his years working on the roads of New Mexico, Huston had already gained a good working knowledge of soils and compaction, but he was always eager to learn something new.) There was an old rock-crusher site on the property, and Huston brought a couple of loads of crusher fines (a by-product of rock crushing) up to the building site. The fines were used alone in some walls, the original subsoil mined near the site in other walls, and a mixture of the two in still others. All three soils made good, strong walls.

As the material for the walls soaked up water in the stockpile, the Hustons poured the foundation. Since there would be more than 365 tons of earth in the walls of the house, the foundation had to be substantial. Footings for all the earth walls were dug 20 in. below grade. For the exterior walls, the footings are 38 in. wide by 8 in. thick with four pieces of ½-in.-dia. rebar running along the bottom; interior walls have a 20-in. by 8-in. footing with two pieces of ½-in.-dia. rebar. The foundation is 26 in. high from the bottom of the footing trench to the bottom of the earth walls. The earth walls sit on a 6-in.-thick stem wall that is 34 in. wide for the outside walls and 16 in. wide for the interior walls (in both cases the width of the raw earth walls). With high-mass, heavy walls, a footing that is the same width as the earth walls is often adequate, but in this case the weight of the house and the nature of the building material dictated the necessity of a greater bearing surface. Huston stepped the foundation above the footing as is typically done, but also added a second step right at floor level to provide a screed line for the slab floor and the outside concrete apron that surrounds the house. This lip is similar to the one used in the Easton system for supporting the form boards (Chapter 10)—on the Huston job, the inside form panels rested on the slab, and the outside panels rested on the lip.

The entire foundation was poured at the same time. The interior slab was poured before the walls were rammed. The concrete apron was poured after the house was plastered.

Like the Sturgeons in Chapter 10, Huston also installed two cooling loops in the house, one on the east side and one on the west. Each is made from 18-in.-dia. duct pipe with galvanized boxes for an air intake and outlet at both ends. These were concreted in before the slab was poured. In the east wing, the intake is in the kitchen floor under a grate and the outlet is in the master bedroom just under the windows. In the west wing, the intake is on the north wall of the guest bathroom and the outlet is under the windows on the south wall of the guest bedroom. The loop is completed by an overhead duct made of sheet metal, which runs between the vigas and connects the same pairs of rooms as the duct pipe does. Mary Huston doesn't like the inlet in the kitchen because it's too hard to clean and it is too easy for things to fall through the grate. Don Huston says he knows that the system moves air around, but he is not sure that it works well enough to be worth the extra work and expense of installation.

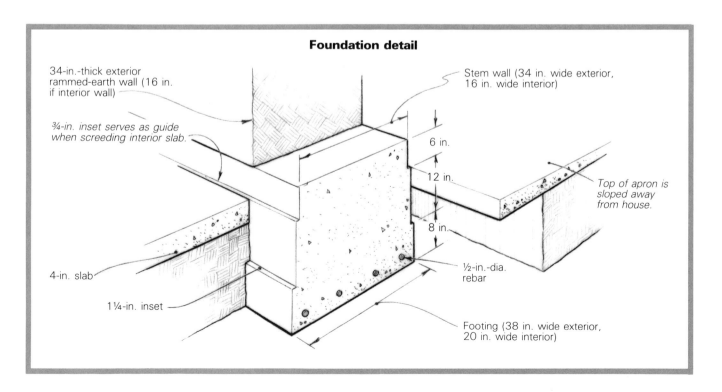

Foundation detail

34-in.-thick exterior rammed-earth wall (16 in. if interior wall)

¾-in. inset serves as guide when screeding interior slab.

Stem wall (34 in. wide exterior, 16 in. wide interior)

6 in.

12 in.

8 in.

Top of apron is sloped away from house.

4-in. slab

1¼-in. inset

½-in.-dia. rebar

Footing (38 in. wide exterior, 20 in. wide interior)

Forming the walls

It is really the method of forming that distinguishes the rammed-earth builders in this book from each other. The Easton form is probably closest to traditional earth-forming methods, where one set of forms was moved around to create all the walls. Builders like the Hustons, who use a truckful of modified concrete-pouring forms, form up large sections of wall at one time. To a certain extent, the forms determine what the house will look like—Easton's module lends itself to single earth panels surrounded by wood trim, while Huston creates huge monolithic walls without a break.

As I discussed in Chapter 10, the basic problem in forming concrete or rammed earth is to create a form that will withstand the weight and pressure of the material and yet be light enough to load, unload and set in place easily. Traditional rammed-earth forms were made of planks and bracing. Some builders, such as Easton, concentrated on portability and ease of setup, while others looked to manufactured systems designed for concrete pouring, where entire buildings were formed at one time. The rammed-earth builders who use modified concrete-pouring systems have invested a great deal of money in forms and approach the issue from a purely dollars-and-cents point of view—they look at forming in terms of maximum volume of wall per man-hour of labor.

Rammed-earth builders found a variety of steel and plywood concrete-pouring systems on the market. Many of these used standard 2x4 panels with narrower sections available for end boards and to take up the slack in walls that weren't built on a 2-ft. module. These panels were fastened to each other with wedges or pins that went through the steel edge of one into the edge of the adjacent one, and were named according to the combination of wedges and pins used. Of course, there also had to be a way to tie the forms together from opposite sides of the wall while maintaining a constant wall width. The ties had to be strong enough to withstand the pressure that would try to push the panels outward. For pouring concrete, metal form ties are used, fastened with pins and wedges at the point where two panels come together. Steel ties are required on 2-ft. centers horizontally and on 2-ft. or 3-ft. centers vertically. This tie system works well for concrete, but when you're standing in one of these forms trying to ram earth with a bucking backfill tamper, the steel ties get in the way. Also, lifting a 75-lb. (or more) tamper over these ties every 2 ft. is tiring. The Hustons figured out a way to use a concrete-forming system and at the same time to minimize the interference of the ties.

The Hustons made their own plywood-and-steel reinforced panels and used all the hardware that is used for concrete-pouring forms, but they made their panels 4 ft. wide. This way the ties, which are fastened at the ends of the panels, are on 4-ft. centers. The wider panel is braced with steel for extra strength and 2x4s (whalers) are used across the outside of the panels to keep the forms in line. These whalers are supported by the extending ends of the form ties and are clipped in place with the same type of wedge bolt that holds the forms together. (The whalers also provide a convenient ladder for climbing into and out of the forms.) Once the forms in any concrete-forming system are removed, the metal form ties, which stick out of the walls, must be pulled out or cut off.

The Hustons' forms are made of 4x4 panels of ¾-in.-thick plywood braced with 2-in. channel iron around all the sides. On each panel there are also two strips of channel iron spaced 15 in. apart and centered on the plywood. The forms butt against each other along the foundation wall and are kept in line with 2x4 whalers secured by the form ties. Photo by Stan Huston.

The forms are stacked for an 8-ft.-high wall, and the Hustons built enough forms to pound 100 running ft. of 8-ft.-high wall at a time. One advantage of concrete-pouring forming systems is that there is no problem with corners or corner forms. Wall forms are locked together at the corners and walls are rammed continuously without a break. With old-fashioned slip forms or with the method Easton used to build the Sturgeons' house, corners present special problems. Some early builders in this country simply designed their structures with windows and/or doors at the corners so they didn't have to deal with corner forms. Easton's solution in the Sturgeon house was to pour concrete corners. The temptation is to simply run each wall up to the corner and butt the walls against each other, but corners built this way are not really attached, and if there is any movement in the walls, the corner will open up. In an adobe house this problem is solved by overlapping the bricks at the corners, and in a stone house with the use of reinforcing bar that runs through the corner from both walls.

In the Hustons' house, the formwork is run along the walls and around the corners—the corners are therefore a part of the walls. Where the forms turn the corner, an extra section of form is needed on the outside to allow for the thickness of the wall. Visualize a 12-ft. by 12-ft. room formed up with Huston's form system. With 3-ft.-thick walls, the outside wall length will be 18 ft. and the inside will be 12 ft. With 4x4 panels, it will take three panels to form the inside and four standard panels plus a 2-ft.-wide panel for the outside. Alternatively, the outside could be done with three panels plus a 3-ft. section at each corner. (The latter method ensures that the form-tie connections between the panels will line up between the inside and outside walls.)

For the rounded corners on the outside walls of the house, the Hustons used quarter-piece lengths of Sonotube. Huston laid the pieces of it into the proper corner of each form, pushed a little dirt around them to hold them in place, then rammed.

The Hustons began by setting up the forms for the interior walls on the slab up against the 6-in.-thick stem wall. The garage walls were rammed second, and the exterior walls third. The slab gave an even surface for the Bobcat front-loading tractor to work on. Even the outside walls were loaded from the inside. The Hustons used two Ingersoll Rand 341 backfill tampers hooked up to a 125 Schramm compressor (125 cfm, 110 psi). Huston was no stranger to the backfill tamper—he had used them in his highway work for the last 30 years. State highway regulations call for 100% soil compaction around culvert pipes, and for getting that first batch of dirt compacted around the bottom there's nothing more versatile than a pneumatic backfill tamper.

A crew of four worked the ramming process. Stan Huston ran the loader, and Don Huston spread the soil evenly in the forms with a shovel for the two hired crew members who ran the tampers. The soil was loaded into the forms in 8-in. layers, and pounded to 4½ in. to 5 in. The bottom row of forms was set for a given wall section, filled with soil and then rammed until the forms were three-quarters full. Then the top layer of forms was stacked and the walls rammed to their full 8-ft. height.

For doorways, the Hustons simply ended a wall and left a space. Here, and where the freestanding interior walls ended, end boards were attached to the forms with the same wedge bolts used to tie together the rest of the system. End boards were the same as the wall panels but narrower—34 in. for the exterior walls and 16 in. for the interior. The bond beam would later provide the doorway lintels (p. 132).

Window openings were run to the top of the wall, also to be topped by the bond beam. Cross-braced rough bucks, set on leveled earth and plumbed, were placed in the forms when the earth wall reached the height of the bottom of the sills. The Hustons used one set of blockouts with 34-in.-wide sides for all the windows, varying window size by changing the length of the bracing. The sides were long enough to be used for doorways and were run wild over the tops of the forms on short windows.

Earth builders handle electrical runs in a variety of ways. In the Southwest, codes allow for the cable to be rammed directly into the wall. Some builders ram conduit into the walls and then thread wiring through it. Others scratch channels into the green earth and run either conduit or shielded cable through them. The latter are later plastered over. Some builders use a combination of techniques. The Hustons tried to put as much wiring as possible into the interior stud walls, but where they needed an outlet box in an earth wall they either rammed direct burial cable or used plastic conduit. Boxes rammed into the walls were wired to the form face through holes drilled in the plywood. The direct burial cable was laid in the bottom of the forms, then carefully pulled up into the first lift of dirt. The trick was to remember where the electrical boxes were wired when it was time to strip the forms. If the tie wires weren't snipped before the forms were pulled down, the electrical box would be pulled right out of the wall. The plumbing all comes through the slab and is hidden in cabinets or stud walls. A few vent pipes are rammed into the earth walls.

The niches in the living-room wall that display the Hustons' collection of Indian art were created with 8-in.-thick hunks of expanded polystyrene rough-shaped to the size of the niche and then wired to the form.

After the walls were rammed and the forms stripped, the Hustons removed the form ties with a Dee stake puller, which enabled them to reuse the ties. Other builders using the form-tie method usually just cut the ties off with a torch.

The Bobcat loads dirt in the form. Photo by Stan Huston.

The end board in the forms is held together by angle iron. The 2x4 whalers are attached at the form ties.

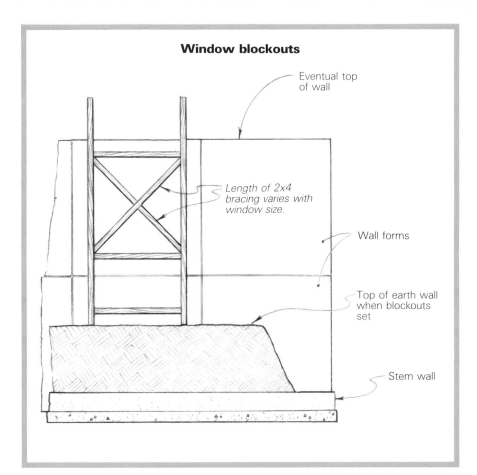

Window blockouts

Eventual top of wall

Length of 2x4 bracing varies with window size.

Wall forms

Top of earth wall when blockouts set

Stem wall

Here the forms are being stripped, and the ties are ready to be pulled out. Each 4x4 section has two ties on each side, 4 ft. on center. Photo by Stan Huston.

Here all the walls are rammed. Each wall sits on a 6-in.-thick concrete stem wall. The dark-colored walls are made from soil mined on the site. The light-colored walls are from crusher fines found at an old rock-crushing operation on the property. The medium-colored walls are a mixture of soil and the crusher fines. Even though these walls were unstabilized, they have eventually hardened to a rock-like consistency. Photo by Stan Huston.

The holes were touched up when the forms were removed. Round corners were also shaped up with an ax as soon as the forms were removed. It was important to do the touch-up work quickly, because the walls started to harden right away: the crew bent a lot of nails putting on the insulation and stucco wire for the plaster.

The Hustons' walls were unstabilized, but even so have hardened to a rock-like consistency. When I stayed with the Hustons in the spring of 1985, I tried scraping an unplastered section of wall in the garage with the steel blade of my knife and couldn't even dent it. The fact that an unstabilized rammed-earth wall will harden and even increase in strength for a long time should be no secret to anyone who has roamed through the literature. Ralph Patty, in a widely quoted study printed in the *Engineering News Record* in 1936, found increases in strength in rammed-earth walls of 33.7% at one year and 45% at two years over his figures for six months.

The chamfers on the corners of this doorway have been rounded with an ax while the earth was still green. The seam lines are from the form boards, and the walls show good compaction. Photo by Stan Huston.

The bond-beam pour has begun. The bond beam is 6 in. thick except over door and window openings, where it is 12 in. thick for a sturdier lintel. The tie wires are visible on top of the earth wall, as are the two pieces of rebar. Threaded bolt has been put into the wet concrete to tie down the vigas. Photo by Stan Huston.

The bond beam

Once the walls had been completed, the Hustons started forming the bond beam. In this house, the bond beam is also the headers for the doors and windows. The Hustons used 8-ft.-long by 12-in.-wide pieces of ¾-in.-thick plywood to form the 6-in.-thick beam. At door and window openings, the thickness was increased to 12 in. to give a sturdier lintel and to lower the headers of the windows and doors. (Where the bond-beam formwork crosses an opening, however, a bottom must be built for the forms and bracing put in underneath during the concrete pour.) All the headers are trimmed in weathered boards taken from an old shed on the property. The boards were placed in the bond-beam formwork on the bottom and against both sides. The sides of the boards that would be against the concrete were spiked with 16d nails—the nail heads protruding into the concrete ensured a firm anchor. Two runs of ½-in.-dia. rebar were laid in along the tops of the walls, and six pieces over the door and window openings. Spreaders made from 1x2s were nailed to the top of the forms to keep the sides from bowing, and the forms were wired together with #9 wire just above the earth at the tops of the walls.

The formed-up bond beam was then poured with concrete and the threaded bolts that would secure the vigas were set into the wet mix. Once the concrete had cured, the forms were stripped. The bond beam would later be plastered.

The vigas and roofing system

In the two wings of the house on either side of the living room, the ceilings are 8 ft. high. The ceiling of the living room is 12 ft. high. The top 4 ft. of the living room is framed with 2x6s on the east and west walls, and then a double stud wall is framed on top of the north wall for extra storage. Where the walls are attached to rammed earth, 6-in.-long screws make the connection. (Wherever there is a connection to the earth walls—for doorjambs, windows or framed divider walls—the same screws were used.)

The log rafters, called vigas, were laid down for the roofs of the wings before the 4 ft. of stud wall was added to the living room. The upper frame walls were built on top of the vigas that were used in the wings.

Large vigas (going left to right) carry the weight of the framed wall and living-room roof above it. The upper vigas are supported by the 2x6 framing and extend beyond the windows. Photo by Stan Huston.

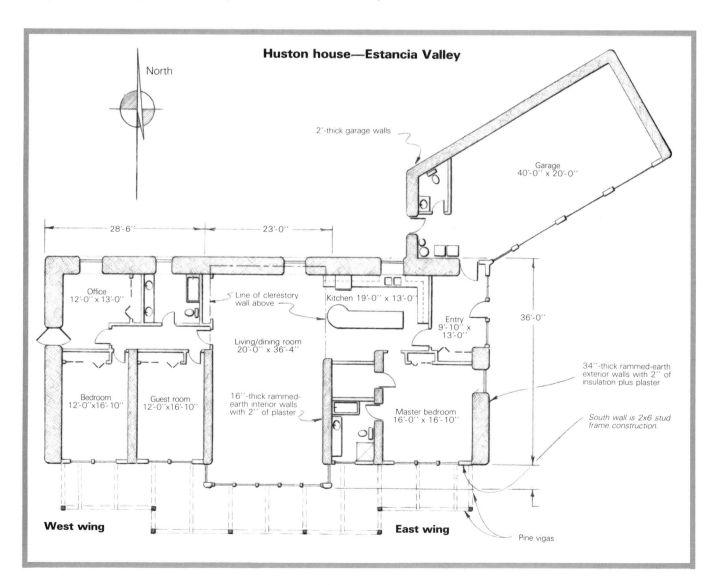

Huston house—Estancia Valley

North

2'-thick garage walls

Garage
40'-0'' x 20'-0''

28'-6''

23'-0''

Office
12'-0'' x 13'-0''

Line of clerestory
wall above

Kitchen 19'-0'' x 13'-0''

Entry
9'-10'' x
13'-0''

36'-0''

Living/dining room
20'-0'' x 36'-4''

34''-thick rammed-earth
exterior walls with 2'' of
insulation plus plaster

Bedroom
12'-0''x16'-10''

Guest room
12'-0''x16'-10''

16''-thick rammed-
earth interior walls
with 2'' of plaster

Master bedroom
16'-0'' x 16'-10''

South wall is 2x6 stud
frame construction.

West wing

East wing

Pine vigas

For handling the vigas, a forklift attachment on the Bobcat was used. In the wings of the house, the vigas were lifted to the tops of the walls and the crew rolled them into place on 3-ft. 3-in. centers. In the living room, the 29-ft.-long vigas were lifted in place with a winch truck and 20-ft. boom.

The vigas in each wing run two different ways. Most of them are 8 in. in diameter and 21 ft. long, except for the first one over the west end of the kitchen, which acts as a support beam for the living-room wall and roof above and is close to 12 in. in diameter. The threaded anchor rod that was set in the bond beam holds the ends of the vigas in place. First the tops of the vigas were power-planed flat to accept the 2x6 tongue-and-groove roof decking, then the threaded rod was extended with long nuts and extra rod so that it would reach past the tops of the vigas. Holes were drilled in the decking directly over the anchor rod, and nuts and washers were screwed down on top of the decking to hold the vigas firmly in place. In the places where the 2-ft.-on-center anchor rods hit the viga, the rod was run right

through the viga and the decking. The ends of the vigas that rest on top of the stud walls were secured to the top of the walls with metal hurricane straps—the straps were given a half-twist as they came off the vigas to give a flat nailing surface.

The 2x6 framing for the tops of the living-room walls is built up over the tops of the vigas that run north to south across the two wings. The 29-ft.-long vigas that run east to west in the living room are supported by the framing, and extend 3 ft. beyond the walls for the roof overhang. Of the 22 high windows in the living room, eight are operable (for venting) and 14 are fixed. The fixed windows vary in size because the vigas vary in diameter, and the Hustons used the fixed windows to take up the slack.

Once the vigas were in place, the Hustons decked the whole building with 2x6 tongue-and-groove roof decking. The sloping living-room roof was originally slated for glass, but the more Huston thought about it, the more he worried that the house would get too hot. After seeing the way the

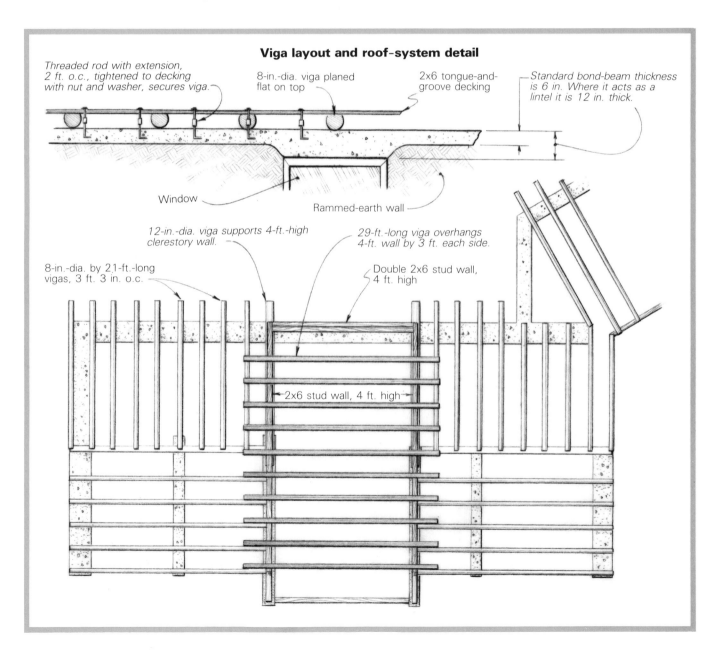

Viga layout and roof-system detail

Threaded rod with extension, 2 ft. o.c., tightened to decking with nut and washer, secures viga.

8-in.-dia. viga planed flat on top

2x6 tongue-and-groove decking

Standard bond-beam thickness is 6 in. Where it acts as a lintel it is 12 in. thick.

Window

Rammed-earth wall

12-in.-dia. viga supports 4-ft.-high clerestory wall.

29-ft.-long viga overhangs 4-ft. wall by 3 ft. each side.

8-in.-dia. by 21-ft.-long vigas, 3 ft. 3 in. o.c.

Double 2x6 stud wall, 4 ft. high

←2x6 stud wall, 4 ft. high→

house performs, he is now glad that he covered this area with the decking instead. (The decking also provides a good base for the panels of the solar hot-water system.)

Huston has only two afterthoughts on this part of the construction. He wishes he had planed the tops of the vigas a little flatter, which would have made setting the decking easier. And he wishes he had let the vigas dry more thoroughly before installing them—as the vigas dried in the house, they twisted and pulled away the plaster. (He probably could have minimized twisting had he sawn a kerf down the top of each viga to relieve tension as it dried—an old log-builder's trick.)

With the decking completed, the Hustons nailed down a layer of ⅝-in.-thick sheetrock for a firestop. On top of that they applied a factory-furnished bond of 6-in.-thick expanded polystyrene and ½-in.-thick plywood, facing the plywood up. On all of the roof except the living-room slope, they built up four layers of fiberglass roofing paper and hot-mop asphalt, then added gravel.

The north wall of the house has a 3-ft. roof overhang.

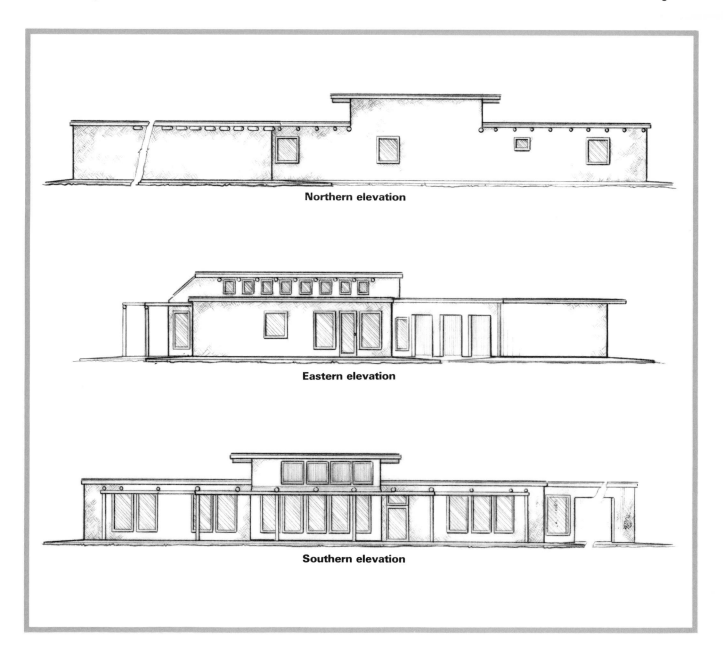

Northern elevation

Eastern elevation

Southern elevation

Finishing up

One thing that the Hustons have in their house besides tons of earth is lots of insulation. Before backfilling the foundation, they nailed 2 in. of expanded polystyrene to all the outside walls from the bottom of the footings to the top of the bond beam. They used 6-in.-long, round-top spiral nails to hold it on and bent quite a few in the curing, rock-hard earth walls. Some information that Huston had read from the University of Minnesota on insulating foundations suggested that horizontal insulation was more effective than the more common vertical application. Just to be sure, Huston did it both ways. The insulation on the walls already covered the vertical face of the foundation, and Huston added the horizontal element by laying out more insulation around the outside of the house. He laid it flat and then poured the concrete apron right on top.

With the insulation on, the house was almost ready for plastering—the south wall just had to be framed and glazed first. Windows along the entire south wall are sealed, double-pane replacement units for sliding glass doors and have vents underneath them. In the evening, the Hustons open the vents and the operable clerestories in the living room to vent the house. Each of the fixed 46-in. by 76-in. panes is held in place with 2x2 redwood stock; the 46-in. by 8-in. vents are screened and each has an insulated door. Rex Roberts proposed a similar system in his book, *Your Engineered House*. He thought that windows should provide illumination and not ventilation, but he proposed that the vents be low on the north wall for air intake and high on the south wall for air outflow.

The Hustons' house uses two different plaster programs—the outside is a three-coat job of cement stucco with what is called a brown coat and a color coat. The interior is a two-coat plaster job with a brown coat for the first layer and a finish coat of hard, white gypsum plaster.

The outside plastering was begun as soon as the south wall was finished and covered with Celotex (an asphalt-impregnated building board). The whole house was then covered with roofing felt, and a layer of stucco wire was nailed on with 8-in.-long galvanized gutter spikes. The layer of cement stucco that goes on the wire is left with a rough surface for the brown coat to adhere to. (Often the cement stucco is applied with a tooth-edged trowel to make a better armature for the brown coat.) The brown coat is the filler and strengthening coat. It is made of fibered gypsum plaster and plaster sand, and is carefully leveled. The color coat, which is only ⅛ in. thick, is made from very fine sand and cement and an oxide coloring agent. To get the sandy texture on this color coat, Huston used a sponge-faced trowel to apply it.

On the interior walls, the brown coat fills in the dings in the earth walls and levels up their surfaces—it is applied thickly, and the fibers keep it from cracking. This coat is usually finished off with a darby, which for some plasterers is simply a 3-ft.-long 1x4 with a 1x2 handle nailed to the back. The plaster is laid on a large area with the trowel and the darby is used to create a perfectly flat and even surface. The finish coat on the interior walls is made from white gypsum plaster and screened fine sand, and is applied after the

brown coat has cured at least a day. The ⅛-in.-thick coat of finish plaster has a slick, hard surface that is tooled with a steel trowel. Some plasterers use a wet brush right in front of the trowel to get this slick surface.

The kitchen, entryway and bathrooms all have ceramic tile on the slab floor. The living room and the bedrooms are carpeted, except for a 17-in.-wide strip of tile in the bedrooms along the south-facing windows, and a 60-in.-wide strip in the living room along the same exposure. This tile gives the winter sun some mass to heat and radiate into the rooms. Huston has plans to build a removable shade for the stylized viga porch on the south side of the house—the shade would be removed in the winter for maximum solar gain and replaced in the summer for maximum cooling.

The interior doors were all made from native New Mexico pine by a door company in Albuquerque. The kitchen cabinets, pantry storage and living-room shelving are all oak, finished with Watco Danish Oil. Watco was also sprayed on the decking, vigas, doors and trim for a clear finish.

Construction on the house had begun in June 1983 and the Hustons moved in during February 1984. I spent some time with the Hustons in the spring of 1985 and felt very comfortable in their house. Even when the outside temperature dropped to 18°F one morning, the gas heaters barely came on. I especially like the entry foyer on the east side of the house. Sealed off from the kitchen by a door, it makes a wonderful sunroom, mudroom, dog room and heat catcher. It was a nice place to have breakfast on a sunny morning and, when the door to the kitchen was left open, it added heat to the house. This room also opens to the garage and is a convenient entrance.

The Hustons find that their house is holding together very well, except where the plaster is breaking around the ends of the vigas. The outside plaster needs repair, too, but Huston knows why. The plasterer went ahead and plastered during a cold spell in spite of Huston's misgivings. The plaster froze while it was curing and now the color coat is failing in several spots.

Huston Construction has already built another rammed-earth house in the Estancia Valley and plans are in the works for several more rammed-earth projects. Is Huston sorry he didn't build with adobe bricks? No, but he's careful not to say anything bad about adobe. Generally, costs are about the same for the two materials in thinner walls, but as you get to 2-ft.- or 3-ft.-thick walls, rammed earth is much cheaper. "How many adobe bricks would it have taken to build this place?" I asked Don. He rolled his eyes back and laughed. "I'd hate to even think about it," he answered, "and that's a fact."

The niches in the walls of the living room house the Hustons' collection of Indian art. They were made by wiring pieces of expanded polystyrene insulation to the forms during ramming.

The kitchen is 20 ft. long and has a 13-ft.-long tiled island running down the middle. Fluorescent fixtures run parallel to the vigas at both ends of the room and a skylight is cut through in the middle.

An 18-in.-thick interior wall divides the master bedroom and master bath. All interior wood doors in the house have been left unpainted.

Albuquerque Adobe

CHAPTER 12

yle Hymer-Thompson built his first adobe house for himself in Redlands, California, with the aid of a general contractor. With its curved windows and inverted arches, the house was a showpiece (see photos, pp. 96 and 140). When it was time to move on to his next inspiration, Hymer-Thompson sold the house and built his second adobe in Cerrilos, New Mexico. This house, too, turned out to be an attention-getter. The third house Hymer-Thompson built, which is in Albuquerque, New Mexico, and is the one I'll focus on in this chapter, is the most modest of the trio. I include it to show that an inexpensive earth house needn't look cheap or take shortcuts.

News of Hymer-Thompson's adobe houses came to me through an article in my college alumni magazine. I was intrigued by the idea that another person with a background similar to mine had ended up building earth houses, and amused at the follow-up letters to the editor that resulted from the publication. An engineer from the class of '72, who wrote that "adobe structures are tested in every earthquake: they fail every time," went on to sing the praises of modern wood-frame houses. An architect responded that after 21 years in Arizona, "I've noticed that adobe has never been used by a qualified architect or engineer, but by the unqualified with emotional motivation." Some of the Arizona architects I know who build with adobe, such as Michael Frerking and Robert Barnes, would probably be amused that the author of this comment works in the field of "development and research."

These negative reactions to earth-building are by no means uncommon, though, and one of the questions that is most frequently asked of the earth builder is, "How do you convince your building department and banker that your earth house meets the requirements of the building code?" Hymer-Thompson has had no problem in New Mexico, where he now lives and works, but the story when he built his first adobe house in southern California was quite different. His struggle is an interesting one, and I'll tell a little about it before going on to the details of the building of the Albuquerque house.

Hymer-Thompson bought his lot in Redlands, California, in 1973 and then shopped for a construction loan. The bankers asked, "Why do you want to build a home out of dirt? Isn't adobe outmoded? Won't it wash away in the first good rain? Why are the rooms circular? How are you going to get windows in these curved rooms? How do we know you can complete this home?" They were worried about the code, resale value and the fact that an owner-builder is a high-risk investment compared to a bonded contractor. Hymer-Thompson was worried about the same things and wasn't about to build a place that wasn't comfortable, that he couldn't finish or that wouldn't sell.

The San Bernardino County Department of Building and Safety had its own version of builder's doomsday. They said he couldn't build because there was a ban on septic systems in the area, that his 1½ acres were zoned for one house on 10 acres, that his adobes would have to be tested at a soils lab, that the house needed reinforced-concrete columns on 2-ft. centers in all the walls, and that he needed more complete drawings. They figured they wouldn't see Hymer-Thompson again.

They were wrong. Hymer-Thompson found that the septic ban was only for condominiums and apartment buildings. A lawyer in his real estate office verified that the land had been legally subdivided, and the local testing lab found that the adobes he was going to use met all county requirements for compression, modulus of rupture, moisture and absorption. A structural engineer demonstrated that reinforced-concrete columns on 4-ft. centers were adequate. And Hymer-Thompson completed six pages of building plans showing all the details.

The exterior of Lyle Hymer-Thompson's 1625-sq.-ft. Albuquerque adobe house may look unspectacular, but behind it is a comfortable and unique interior with 10-in.-thick stabilized walls. This bricked-in area under the latillas is a pleasant spot to sit in the shade and enjoy the view of the Sandia mountains to the southeast.

Hymer-Thompson built his first adobe house in Redlands, California, with the help of a general contractor. The home is made of a series of semicircles and rectangles. Bronze-tinted Plexiglas is used in the arched windows. The adobes are unplastered.

Inside and out, Hymer-Thompson left the adobes unplastered to show off their natural beauty. The south side of the house is mostly glass for solar gain.

Still the bankers stalled and the summer wore on. So Hymer-Thompson started to dig the foundation and septic by hand with the aid of some college students who volunteered to help. (Hymer-Thompson continues to prefer to hand-dig his footing trenches, because the walls in his houses are usually quite close together, and because he likes manual labor better than backhoes.) Finally he hired a contractor who agreed to work with him and his helpers. By providing free labor, Hymer-Thompson was able to keep costs down and learn adobe-laying from the first-class masons whom the contractor brought to the job. Finally, with a bonded contractor on the job, the bank came through with the construction loan.

After waiting nearly two years to get started, Hymer-Thompson and his contractor built a good house. The students learned quickly and worked hard. The experienced masons taught the finer points of mixing mortar and laying adobe. When Hymer-Thompson moved to New Mexico, he got his contractor's license, and now his only problems are the usual ones of any builder.

In the Albuquerque house, Hymer-Thompson made the walls 10 in. thick and used stabilized adobes. After the walls were up, he sprayed the interior surfaces with a silicone sealer to prevent dusting. As in all Hymer-Thompson's houses, the adobes are unplastered inside and out to show off the natural colors and patterns of each one. The house has 1625 sq. ft. of living area and many of the features of Hymer-Thompson's previous houses—an open floor plan and arched windows and entries, for example. It is oriented to the south and glazed for solar gain, and with all the thermal mass in the walls and floor, it is easy to heat and cool. Though its exterior is much plainer than Hymer-Thompson's first two adobes—no sculpted forms or particularly exciting shapes—the house has a certain magic on the inside. The entryway is a rich combination of the weathered wood of the handmade front door (see photo, p. 146) and the earth-colored adobes. The Mexican-style interior brick floors are dry-laid in a sand base and then sealed with linseed oil. An archway frames the bedroom, and the skylight above the bed floods the room with light. The poured-adobe floor of the darkroom, discussed on p. 146, is sealed in traditional fashion with animal blood purchased at a local slaughterhouse. (The blood hardens, and doesn't smell or draw flies.) In some places the surface has peeled and there are some nicks from high-heeled shoes, but overall the floor has been a success. According to Hymer-Thompson's wife, Karen, a photographer who spends a lot of time on her feet, the poured-adobe floor is not nearly as tiring to stand on as concrete would be.

The construction budget for the house was tight— $25,000 for the whole project, including a crew of three but excluding Hymer-Thompson's own labor. (He plans to make his profit when he sells the house; when I last visited him, it was on the market for $85,000.) Breaking down his costs, Hymer-Thompson found that only 6% of his budget was used on the adobes, mortar and stabilizer. He spent another 8% of the tally on the roughsawn lumber and timbers, 11% on labor, and 16% percent on the electrical and plumbing systems.

Hymer-Thompson house—Albuquerque

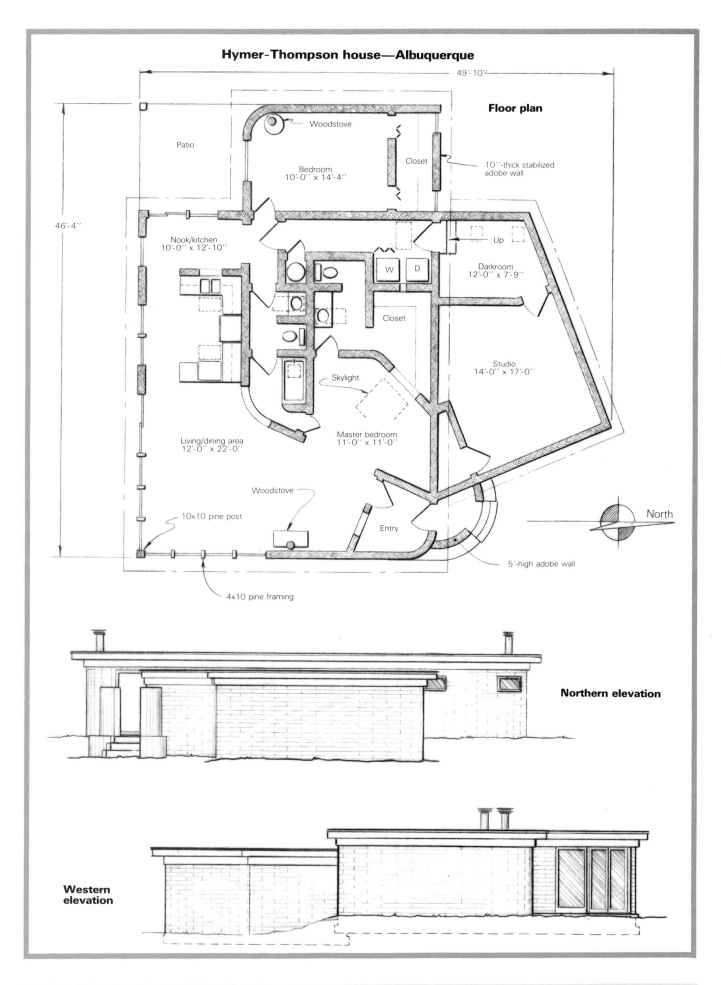

Floor plan

49'-10"

46'-4"

Woodstove

Patio

Bedroom
10'-0" x 14'-4"

Closet

10"-thick stabilized
adobe wall

Nook/kitchen
10'-0" x 12'-10"

Up

W   D

Darkroom
12'-0" x 7'-9"

Closet

Studio
14'-0" x 17'-0"

Skylight

Living/dining area
12'-0" x 22'-0"

Master bedroom
11'-0" x 11'-0"

Woodstove

North

Entry

10x10 pine post

5'-high adobe wall

4x10 pine framing

Northern elevation

Western elevation

The foundation

The more than 6,000 adobe blocks to be used in the 10-in.-thick walls of the house measured 10x4x14 and weighed more than 80 tons. On top of the walls would be a wood bond beam and a flat roof. Hymer-Thompson had to be careful that the footings and foundation would support all of that weight.

All the adobe walls use the same foundation system, except for the wall that divides the studio and darkroom from the rest of the house. Because the house is built on two levels, the finish floor in the darkroom is 2 ft. below the finish floor in the rest of the house. For a little extra engineering, Hymer-Thompson ran a 4-in.-thick by 27-in.-high concrete retaining wall along the inside of the foundation that supports this one wall. (The retaining wall is tied into the footing with ½-in.-dia. rebar.) Therefore, whereas the footings for all the other walls are 12 in. wide (2 in. wider than the walls), the footing for this wall had to be 4 in. wider than the other footings. You might remember that the house discussed in Chapter 10 has footings the same width as the earth walls, and the footings of the house discussed in Chapter 11 are 4 in. wider than the walls. This variation is a result of the relative weight of the house and the bearing strength of the soil the house will be built on. A wider footing distributes the weight of the house over a greater area, so the more stable the soil, the narrower the footings can be. All the footings in Hymer-Thompson's house are reinforced with two runs of ½-in.-dia. rebar.

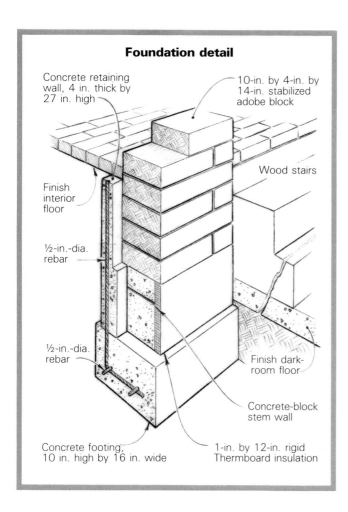

Foundation detail

Concrete retaining wall, 4 in. thick by 27 in. high

10-in. by 4-in. by 14-in. stabilized adobe block

Wood stairs

Finish interior floor

½-in.-dia. rebar

½-in.-dia. rebar

Finish dark-room floor

Concrete-block stem wall

Concrete footing, 10 in. high by 16 in. wide

1-in. by 12-in. rigid Thermboard insulation

Hymer-Thompson prefers to make the foundation walls for his houses from concrete block rather than poured concrete. This is because he likes to insulate them on both sides, and the Thermboard rigid insulation he is able to use on concrete-block stem walls costs about four times less than the Foam-Form insulation needed for a poured foundation. The stem walls of this house are 8 in. wide and are made with two runs of block—an 8-in.-high block on the bottom and a 4-in.-high block on the top. The 1-in.-thick rigid insulation totally covers the block on both sides and brings the stem walls out to 10 in. wide, which matches the width of the adobe walls.

Laying the adobes

After the foundation was built, the first step was to put in the door bucks and the roughsawn pine 4x10 wood sills in the living room that go under the east and south floor-to-ceiling windows. (The other windows would be attached to modified "gringo blocks"—see p. 144.) Hymer-Thompson was then ready to begin laying adobe. He calls his adobe work simplistic, labor-intensive, low-key and low-tech—he hires the same people for some jobs, but usually works with unskilled workers, such as students or homeowners who want to learn the process.

The advantage of buying adobes from an adobe yard is that they are produced to the standards of the Uniform Building Code, and they have been dried. Hymer-Thompson bought his adobes from an adobe yard in the Albuquerque area that delivered them to the site in 200- to 300-block batches. (Each stabilized adobe weighed 25 lb. to 30 lb., compared to the 15 lb. to 20 lb. of an adobe made with straw, so a 300-brick load weighed in at more than four tons.) If the load being delivered on a particular day could be set right where the truck off-loaded it, the driver would do the unloading. But if the adobes were needed inside the building, the crew had to move them to the work spot bucket-brigade style, the first man in the chain tossing the adobe to the next man and so on down the line. As he began the adobe work, Hymer-Thompson also had a load of good adobe soil delivered to the site. From this he would make the mud mortar that would glue the adobes together, and later the poured-adobe floor.

In California, the code requires that adobe builders use a cement-based mortar such as would be used for laying stone or fired bricks. The mortar is unquestionably strong, but because it has a different coefficient of expansion than the adobes, some builders argue that it actually pulls away from the bricks over time. If you use cement mortar with unplastered walls, another problem is that as the softer adobe erodes on the exterior of the house, the harder cement mortar is left standing out in relief. In New Mexico, you can use mud mortar, and the wise builder chooses a good adobe soil with the same makeup as the adobes for this.

Because Hymer-Thompson used stabilized adobes to build his house, he also stabilized the mortar. (On a job where he is not stabilizing, he lets the mortar mix itself by simply running a hose in a small pit dug at the edge of the pile of adobe soil, a method commonly used in Mexico.) In a wheelbarrow he mixed four gallons of water with a cup of emulsified asphalt, then mixed in 15 to 20 shovelfuls of the adobe soil. He used a hoe to stir, and the resulting mortar had a consistency that was thick and sloppy but that wouldn't run off when shoveled onto the adobes.

Hymer-Thompson began the adobe work by laying two courses of adobe around the perimeter of the house and on every interior wall. He shoveled the mortar onto the wall, then spread it about ¾ in. thick with a 12-in. mason's trowel. When laying a 14-in.-long adobe, most builders strive to keep the vertical joints at least 4 in. over from the joint below for strength. Builders such as Hymer-Thompson, who don't plaster their walls, strive for consistent patterns and level horizontal lines.

Hymer-Thompson usually spreads enough mortar to lay about 10 or 12 adobes at a time. His old trowel is not quite the same shape it used to be—the pointed end is worn down to a stub as a result of pointing mortar, and the edge is worn concave from chopping adobes in half or shaping them to fit odd spaces around inserts (such as electrical boxes or plumbing pipes). A well-worn trowel is a badge of experience for the *adobero* (adobe mason).

Once he had completed the first two courses of adobe, Hymer-Thompson put in the electrical boxes. The outlets are set on top of the second course, about 9 in. above the finish floor. Nails are put through the side of each box, but they're not pounded into anything as they would be in a stick house—the mortar that will go around the nails holds the box in place after the mortar hardens. The box is laid in a vertical mortar joint and the corner of one adobe cut off to accept it. The adjacent adobe slips between the nails protruding from the side of the box and butts against the mortared side of the box. The UF cable that Hymer-Thompson generally uses for wiring comes out of the box in the middle of the wall (5 in. in from either side), goes through the first mortar joint and is run up subsequent courses either through a hole drilled in the adobe or through joints.

In New Mexico, the electrical wiring in a commercial building must be run through conduit. Hymer-Thompson's electrician on this job hadn't done any residential structures and so favored using conduit. Hymer-Thompson went along with him, but feels it is much easier to install the UF cable as just described than it is to work with the stiff conduit.

As the adobe work progressed, Hymer-Thompson concentrated on finishing one section of wall at a time—it's demoralizing to work on the whole house at once because you see very little progress. But if you stick with one section, you become familiar with the electrical and plumbing systems and the window placement, and so can work more efficiently. Hymer-Thompson built up the corners first (adobes are overlapped alternately from each wall that meets at a corner), then worked up around window and door openings. The resulting straight stretches in between could then be laid up quickly.

Some builders make corners out of two pieces of lumber for story poles, then plumb and brace them to the foundation on all the corners. Each course of adobes is marked on the pole, and by watching the marks, the *adobero* is able to keep the walls level and plumb. Other builders use a string line that is raised for each course and held against the verti-

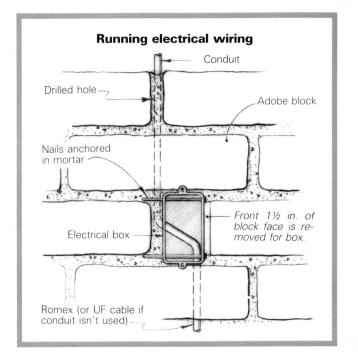

Running electrical wiring

Conduit

Drilled hole

Adobe block

Nails anchored in mortar

Front 1½ in. of block face is removed for box.

Electrical box

Romex (or UF cable if conduit isn't used)

The door bucks are put in place before the walls go up. The electrical conduit is run through holes drilled in the adobes. Photo by Chuck Gruver.

The corners of the back bedroom have been laid up first, which allows the straight section in between to be laid up quickly. Photo by Chuck Gruver.

cal surface of the wall to keep it plumb. Hymer-Thompson says he originally used strings and plumb bobs, but now he just depends on his eye and his carpenter's level.

Most of the plumbing in the house is run through the walls. The adobes are simply carved out with the trowel to go around vent pipes and water lines. In the bathroom next to the living room, some of the water lines were installed after the walls were up; these were run in the mortar joint on top of the first course of adobes and then remortared.

Obviously, the work slows where you have to carve or drill the adobes. It also slows at the window and door bucks. Window bucks in the middle of a wall are handled the same as door bucks, except that they are laid on a bed of mortar and braced on the wall. As the walls are laid up, nailers must be put in for the bucks to attach to. These nailers, called "gringo blocks," are generally 4-in.-high wood blocks mortared right in with the adobes. Normally they're later hidden by the plaster, but because Hymer-Thompson doesn't plaster, he uses gringo blocks made of pieces of 1x6 and hides them in the mortar joints. The only problem is that these blocks are harder to find when you're nailing in the bucks, so as soon as he lays one in a wall, Hymer-Thompson drills through from the buck to make holes for 4-in. wood screws.

Since he doesn't plaster, Hymer-Thompson also has to carefully fill all head (vertical) joints and bed (horizontal) joints with mortar as he works. Then he smooths the mortar with the trowel. When the roof is on, he smooths the joints one last time with a wet burlap sack to eliminate any little cracks or flaws that might have occurred during drying. In houses that are to be plastered, the adobes are usually laid crookedly and the mortar has so many gaps in it that uninitiated observers often wonder what is holding the adobes together. The idea, of course, is that a rough surface with lots of voids is going to provide a better armature for the first coat of plaster, which will key into all the openings in the joints.

When the weather is really hot, Hymer-Thompson uses a watery mortar so it will stay workable. If the weather changes abruptly, he just moves to another part of the house to give the mortar time to cure. Generally, he works only four or five courses at a time—in hot weather, the first four courses laid in the morning will be ready for another four or five courses in the afternoon.

The wood "gringo blocks" of this house have been mortared into the wall to be used as nailers and will be plastered over.

Building the arches and circular windows

Hymer-Thompson begins this work by making the formwork. For an arch with, for example, a diameter of 6 ft., he'll use two 3x6 pieces of ⅝-in.-thick plywood or ⅛-in.-thick Masonite; ½-in.-thick plywood is adequate if the formwork won't be used for more than a few arches. He ties one end of a string to a nail driven in at the halfway mark of the long edge of the plywood or Masonite, and the other end to a pencil—he then swings the arch on both pieces of material and cuts out the shapes. Hymer-Thompson then nails some pieces of 2x6 or 2x8 between the arches; for an arch with a diameter of 6 ft., five boards are sufficient. The longest board, which is 3 ft., is nailed in the middle; the other boards are spaced evenly apart and nailed in place. The same basic idea can be used to make the formwork for circular windows.

The form is then positioned and the adobes laid on. Masonite or thin metal can be nailed to the top of the form to close in the spaces, but the adobes are well supported by the edges of the form. Some beginning builders plot each adobe and mortar joint along the edge of the form so that they will come out with an even number of adobes, but Hymer-Thompson doesn't bother to do this. By adjusting the joints as he nears the end of the arch, he can make the arch come out pretty close to the way he wants it—if he should come up with an odd-sized opening, he just cuts a block to fit. The formwork should be left in place for at least a week before it is knocked free.

In addition to the formwork described above, Hymer-Thompson also uses redwood bender board—a thin, flexible board used for defining round-shaped planting areas in a landscape scheme. The form used to make the bedroom entry, for example, was made from a piece of 8-in.-wide bender board. Set in place on both sides of the entry, the board bent to a natural arch. (A longer board will give a fuller arch; a shorter board a flatter one.) To lay up the arch, Hymer-Thompson first cut the adobes in half so that they measured 10x4x7. As he stacked the adobes, they gradually sloped to the center of the arch. He worked from alternate sides to balance the forces and to keep the arch symmetrical, laying up the adjoining walls as he went so they wouldn't be too far behind the arch. (Because the forces in an arch are transmitted to the sides, those adjacent walls are needed for support.) But as he worked his way up, the bender board started to sag, so Hymer-Thompson had to brace it. By the time he was through, the form looked like a giant fan, as you can see in the background of the bottom photo on p. 143. As Hymer-Thompson approached the top of the arch, he left it to chance to determine whether he would have to cut a wedge-shaped key or could get by with a half-block—he says it doesn't seem to matter structurally either way.

To make the round window in the kitchen, Hymer-Thompson cut the adobe blocks in half lengthwise so they measured 5x4x14. He laid them with the long side perpendicular to the wall, so they would protrude 2 in. on each side of the wall and put the circle in relief. To start, he laid a couple of the blocks on the wall for the bottom of the window and put the form in place. Then he laid up some full-size adobes on the wall to support the form, and then some more to make the circle on each side. He worked window and walls together, taking the form in and out and using it more as a guide than as a resting place for the adobes. When he got to the top half of the circle, however, where the adobes needed to be supported as they set up, he left the form in place. He worked slowly, letting the mortar dry hard before going too far.

Although the arch in the curved wall of the living room seems as though it would be complicated to do, Hymer-Thompson found it to be actually fairly simple. He set his level against the 3-ft.-high section of already laid curved wall and laid the adobes in the arch to the level. For formwork, he used Masonite, but he did not nail the two half-circles together. Instead, he made a slot in the top of one and in the bottom of the other, and slipped them together like the cardboard stand for a paper doll. He put them on the curved wall with the end of the flat edge of one piece of Masonite running from the outside of the wall on the left side of the arch to the inside of the wall on the right side of the arch. He ran the other piece of Masonite from the opposite inside to the opposite outside. For this arch he used adobes measuring 10x4x7.

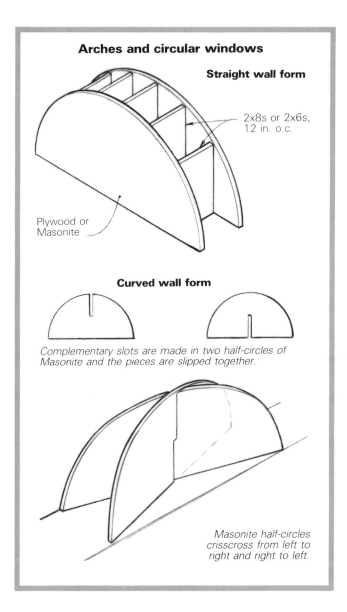

Arches and circular windows

Straight wall form

2x8s or 2x6s, 12 in. o.c.

Plywood or Masonite

Curved wall form

Complementary slots are made in two half-circles of Masonite and the pieces are slipped together.

Masonite half-circles crisscross from left to right and right to left.

Here the bender-board arch form is in place for the bedroom entry. The arch in the background is a wood-backed decorative niche above the bed in the finished bedroom. Photo by Karen Hymer-Thompson.

This circular window in the airlock entry of the hallway is similar to that in the kitchen. Photo by Karen Hymer-Thompson.

Putting in the floors

The poured-adobe floor in the darkroom was an experiment Hymer-Thompson had wanted to try ever since he began building adobe houses. As a frequent traveler and tour leader to Mexico, he had long admired the earth floors found in traditional Mexican adobes.

For this floor, Hymer-Thompson used the same adobe soil that he had used to make his mud mortar. For a base, he simply used the well-tromped soil of the building site. In a wheelbarrow he mixed the soil with water to the consistency of thick yogurt, then he divided the floor area into three 3-ft.-wide sections with form boards and poured it by section. He poured the section on the right first, the section on the left second and the section in the middle last, removing the form boards as he did so. Each section was 4 in. thick, and screeded with Hymer-Thompson's aluminum level. (Poured adobe is so sticky that a board screed such as is used for concrete quickly gums up.) As the floor dried over a period of about a week, it cracked like mad—in some places the cracks were ¼ in. wide. Hymer-Thompson then used a slurry of screened soil (no gravel or little rocks) as a finish coat to fill the cracks. This coat cracked too, but the largest fissures were no more than ⅛ in. wide; most were much narrower. So he poured one more layer of screened-soil slurry about 1 in. thick and ended up with only a few hairline cracks.

Hymer-Thompson had read somewhere about a process for finishing floors with ox blood, but because he couldn't find any of this, he settled for two 5-gal. cans of cow's blood from the local slaughterhouse—for a total cost of $3. He had planned to spray on the blood, but it started to coagulate and he ended up working it in with his hands. He was surprised and pleased that it didn't smell or draw flies and animals. Today there are parts of the floor that are flaking, but Hymer-Thompson feels that this is because he didn't wet the surface of the floor before he added the blood. Consequently, the blood made a hard surface, but didn't bond with the earth. He says that for the next floor he does he is going to stabilize the soil mix with emulsified asphalt and then use a commercially made floor sealer instead of the blood.

The rest of the house is floored with brick dry-laid on a sand base. This is a popular floor in adobe houses and fairly easy to install. Hymer-Thompson laid down a sand base about 3 in. deep and leveled it by dragging his 4-ft. level over it while watching the bubble. He set the bricks in a simple pattern, brick over joint, as they might appear in a wall. After tapping each brick lightly with a rubber mallet to seat it in the sand, he swept sand into the cracks to lock the bricks in place. He then sealed the floor with several coats of linseed oil.

A curved wall in the living/dining area divides this space from the master bedroom and the extra bathroom. Tinted Plexiglas lends privacy to the tub area.

Hymer-Thompson made the front door of his home from 2x6 tongue-and-groove pine. Mexican ceramic masks are used throughout the house as lighting fixtures.

A 10x10 pine post holds up the southeast corner of the roof. The 6x10 bond beam ties the walls together at the top, and the 4x10 sill plate is on the bottom. The 34x76 windows are divided by pieces of roughsawn 4x10 pine.

The bond beam and finishing up

When the walls were at their full height of 7 ft., the 6x10 roughsawn-pine bond beam was attached along the top of all the exterior walls. Holes were drilled in the beam for the conduit, and every 4 ft. a ¼-in.-dia. hole was drilled through the beam and into the top of the adobe wall for the 12-in.-long spikes that would tie the beam to the wall.

After this was done, the east and south walls of the living room, which are mostly glass, were framed. (Remember that before laying up any adobes, pine sills were bolted to the stem wall for these floor-to-ceiling windows.) A 10x10 pine post set in the southeast corner and vertical 4x10s define the windows and sliding door. Windows are double-pane sliding-door blanks of tempered glass measuring ³⁄₁₆x34x76. The bond beam ties the window area into the adobe walls. Interior headers for doorways and hallways are of the same roughsawn 4x10 pine, and strips of 1x pine hold the fixed glass windows in place. All of the exposed timbers and the bond beam are sealed with log oil, which is basically a varnish base with linseed oil for body and paint thinner for penetration. (Some people thin it further with mineral spirits or diesel fuel.) The oil has left the wood a rich yellow color that contrasts beautifully with the earth walls.

The ceilings in Hymer-Thompson's house are just a little over 7 ft. high; some people might find this cramped, but I found it cozy. Hymer-Thompson wanted them low because he felt it would be more energy-efficient. The flat roof is built on 2x8 pine joists set on the bond beam on 2-ft. centers.

The finish ceiling is 1x12 pine. The roof is 90-lb. roll roofing on top of 15-lb. felt applied and sealed with plastic roofing cement.

On the roof is an evaporative cooler that is used four to six weeks out of the year. The rest of the time the house stays cool on its own and the sliding glass doors provide plenty of cross ventilation. In the living room there is a woodstove with a small, thermostatically controlled blower. In the cold weather this stove heats the master bedroom, living room, dining room and kitchen with two or three minutes of blower time. (Hymer-Thompson uses about a cord and a half of wood a season.) Some mornings the solar heating is enough and the Hymer-Thompsons don't even start the stove. There are electric backup heaters under the cabinets in the kitchen, but they've never been used.

As the house was being finished, the garage was closed in for Karen Hymer-Thompson's studio and darkroom. The door to the studio is exactly the same as the front door, and Hymer-Thompson built a modified zia (the stylized sun that has become New Mexico's official logo) out of glass blocks into the wall where the garage doors should have been. In the bathrooms and kitchen, the counters are finished with Mexican tile. The tinted Plexiglas used in Hymer-Thompson's first house (shown on p. 140) makes another appearance in the arched window between the living room and the bathtub—the tint keeps anyone from seeing in, but from a warm tub on a nice evening you can see the city lights and the moon overhead.

Bibliography

Stone construction

Flagg, Ernest. **Small Homes, Their Economic Design and Construction.** New York: Charles Scribner's, 1921.

Kern, Ken, Steve Magers and Lou Penfield. **The Owner Builder's Guide to Stone Masonry.** North Fork, California: Owner Builder Publications, 1976. (Distributed by Charles Scribner's Sons, N.Y.)

McRaven, Charles. **Building with Stone.** New York: Lippincott and Crowell, 1980.

Nearing, Helen and Scott. **Living the Good Life.** New York: Schocken Books, 1970.

Peters, Frazier. **Pour Yourself a House.** New York: McGraw Hill Book Co., 1949.

Schwenke, Karl and Sue. **Build Your Own Stone House Using the Easy, Slipform Method.** Charlotte, Vermont: Garden Way Publishing, 1975.

Watson, Lewis and Sharon. **How to Build a Low-Cost House of Stone.** Sweet, Idaho: Stonehouse Publications, 1978.

Log construction

Building a Log Cabin in Alaska. College, Alaska: University of Alaska Cooperative Extension Service, 1971.

Fine Homebuilding Construction Techniques. Newtown, Conn.: The Taunton Press, 1984. (Contains articles on earth, stone and log building.)

Langsner, Drew. **A Logbuilder's Handbook.** Emmaus, Penn.: Rodale Press, 1982.

Leitch, William. **Hand-Hewn: The Art of Building Your Own Cabin.** San Francisco: Chronicle Books, 1976.

Mackie, B. Allen. **Building with Logs.** B.C., Canada, 1971. (Distributed by Charles Scribner's Sons, N.Y.)

Mackie, B. Allen. **Log House Plans.** New York: Charles Scribner's Sons, 1981.

McRaven, Charles. **Building the Hewn Log House.** Hollister, Mo.: Mountain Publishing Services, 1978.

Earth construction

Adobe and Rammed Earth Housing. United Nations, N.Y.: UN Town and Country Planning Bulletin #4, 1950.

Boudreau, Eugene. **Making the Adobe Brick.** Berkeley, Calif.: Fifth Street Press, 1971. (Distributed by Random House Inc., N.Y.)

Doat et al. **Construire en Terre.** Center for the Research and Application of Earth Technology, Haut-Brie, F-38320 Eybens, Grenoble, France, 1979.

Easton, David. **The Rammed Earth Experience.** Wilseyville, Calif.: Blue Mountain Press, 1981.

How to Build Your Home of Earth. Stillwater, Okla.: Oklahoma Engineering Experiment Station, Oklahoma Agricultural and Mechanical College, Publication 64, 1946.

Kern, Ken. **The Owner-Built Home.** New York: Charles Scribner's Sons, 1972.

Kern, Ken. **The Owner-Built Home Revisited.** North Fork, Calif.: Owner Builder Publications, 1984.

Manual on Stabilized Soil Construction for Housing. United Nations, N.Y.: UN Technical Assistance Program, 1958.

McHenry, Paul Graham Jr. **Adobe—Build It Yourself.** Tucson, Ariz.: The University of Arizona, 1973.

Merrill, Anthony. **The Rammed Earth House.** New York: Harper and Brothers, 1947.

Middleton, G.F. **Build Your Home of Earth.** Milsons Point NWS, Australia: Second Back Row Press, revised 1979.

Miller, Lydia and David. **Manual for Building a Rammed Earth Wall.** Greeley, Colo.: Rammed Earth Institute International, 1980.

Newcomb, Duane. **The Owner-Built Adobe House.** New York: Charles Scribner's Sons, 1980.

Paints and Plasters for Rammed Earth Walls. South Dakota Agricultural Experiment Station, South Dakota State College, Bulletin 336, May 1940.

Rammed Earth Building Construction. South Carolina Engineering Experiment Station, Clemson College, Bulletin #3, revised May 1951.

Rammed Earth Walls. South Dakota Agricultural Experiment Station, South Dakota State College, Circular 149, April 1959.

Rammed Earth Walls for Building. USDA Farmer's Bulletin 1500, revised 1937.

Rammed Earth Walls for Farm Buildings. South Dakota Agricultural Experiment Station, South Dakota State College, Bulletin 277, revised 1945.

Scoggins, H. **The Portalab Manual: Low-Cost Soil-Engineering Tests for Constructing Earthen Buildings.** New Mexico Appropriate Technology Program, funded by USDOE, June 1981.

Suitability of Stabilized Soil for Building Construction. Illinois Engineering Experiment Station, Bulletin 333, December 1941.

Index

Publisher, Books: Leslie Carola
Editor: Laura Cehanowicz Tringali
Design Director: Roger Barnes
Associate Art Director: Heather Brine Lambert
Illustrations: Heather Brine Lambert
Manager of Production Services: Gary Mancini
Coordinator of Production Services: Dave DeFeo
System Operator: Claudia Blake Applegate
Production Assistants: Mark Coleman, Deborah Cooper, Dinah George
Pasteup: Marty Higham, Mary Ann Snieckus

Typeface: Univers, 9½ point
Paper: Maker's Matte, 70 lb.
Printer and Binder: Kingsport Press, Kingsport, Tennessee